# SpringerBriefs in Mathematics





**SpringerBriefs in Mathematics** showcases expositions in all areas of mathematics and applied mathematics. Manuscripts presenting new results or a single new result in a classical field, new field, or an emerging topic, applications, or bridges between new results and already published works, are encouraged. The series is intended for mathematicians and applied mathematicians.

For further volumes:
http://www.springer.com/series/10030

# BCAM SpringerBriefs



**BCAM** ***SpringerBriefs*** aims to publish contributions in the following disciplines: Applied Mathematics, Finance, Statistics and Computer Science. BCAM has appointed an Editorial Board, who evaluate and review proposals.

Typical topics include: a timely report of state-of-the-art analytical techniques, bridge between new research results published in journal articles and a contextual literature review, a snapshot of a hot or emerging topic, a presentation of core concepts that students must understand in order to make independent contributions.

Please submit your proposal to the Editorial Board or to Francesca Bonadei, Executive Editor Mathematics, Statistics, and Engineering: francesca.bonadei@springer.com

(bcam)
basque center for applied mathematics

Aurora Marica • Enrique Zuazua

# Symmetric Discontinuous Galerkin Methods for 1 – D Waves

## Fourier Analysis, Propagation, Observability, and Applications

basque center for applied mathematics

Springer

Aurora Marica
Institute of Mathematics
and Scientific Computing
University of Graz
Graz, Austria

Enrique Zuazua
BCAM-Basque Center
for Applied Mathematics
Ikerbasque, Bilbao
Basque Country, Spain

ISSN 2191-8198 ISSN 2191-8201 (electronic)
ISBN 978-1-4614-5810-4 ISBN 978-1-4614-5811-1 (eBook)
DOI 10.1007/978-1-4614-5811-1
Springer New York Heidelberg Dordrecht London

Library of Congress Control Number: 2014932413

Mathematics Subject Classification (2010): 65T50, 65F15, 70H05, 70H08, 78A05, 78M10, 93B05, 93B07, 93C05, 93C15



Printed on acid-free paper

Springer is part of Springer Science+Business Media (www.springer.com)

*To the memory of Neli-Mioara and Juan Luis*

# Foreword

When, some years ago, I was working on the Preface of the reference [32] of this book, I wrote

"As everyone knows in the Scientific Community, and elsewhere, Jacques-Louis Lions passed away in June 2001, while still active scientifically. He largely contributed in making the *Control of Distributed Parameter Systems* a most important field where sophisticated mathematical and computational techniques meet with advanced applications."

With J.L. Lions' untimely departure, the control of distributed parameter systems (i.e., the control of systems modeled by partial differential equations) had been deserted by its uncontested leader. Indeed, Jacques-Louis dominated this scientific field for more than 30 years by his results, his new ideas, his permanent quest for new problems and novel applications, his outstanding ability at motivating other scientists, and his exciting way of lecturing on difficult topics. To tell the truth, I was doubtful that a leader of comparable dimension will appear soon. Fortunately, a small number of outstanding applied mathematicians proved me wrong (most of them mentioned in the list of references of the present volume), among them *Enrique Zuazua*. In very few years, he produced many seminal publications ([70,71] in particular), was invited to lecture on distributed control related topics at ICM 2006 (ICM: International Congress of Mathematicians), supervised Ph.D. students and postdoctoral collaborators from Spain, France, and other countries, and, very important to me, started to investigate the *computational aspects* of the control of distributed parameter systems. On a personal note, let me say that I was delighted when E. Zuazua and collaborators proved, in the mid-2000s, via a mathematical tour de force, the convergence of a two-grid method I had introduced in the early 1990s, for the exact boundary controllability of the linear wave equation by the Hilbert uniqueness method of J.L. Lions.

The present book, albeit relatively short, is another outstanding contribution to the control of distributed parameter systems and related topics. In this book, E. Zuazua and his collaborator *Aurora Marica* have reported the results of their investigations on the *approximation of the one-dimensional wave equation by symmetric discontinuous Galerkin methods.* More precisely, in this book, the authors discuss how accurately these discontinuous Galerkin methods approximate the propagation and observability properties of the original fully continuous model. Indeed, it has to be realized that some numerical methods which can be used to solve accurately partial differential equations may be badly suited to solve numerically inverse problems (control problems in particular) associated with the same equations. It is precisely the issue that E. Zuazua and A. Marica analyze in this book: namely, show that a particular class of discontinuous Galerkin methods is well suited to solve accurately not only the one-dimensional linear wave equation but also control problems associated with this equation.

This book is a wonderful blending of outstanding partial differential equation analysis and numerics, not encountered together elsewhere. It should interest control theoreticians and practitioners and, beyond control, PDE specialists from both the theoretical and computational points of view.

The authors should be congratulated to have produced in so few pages a most remarkable and exciting piece of work. Actually, I find also remarkable and highly symbolic that this book was completed while E. Zuazua was visiting the Jacques-Louis Lions Laboratory at University P. & M. Curie in Paris.

Houston, Texas, USA
November 20, 2013

Roland Glowinski

# Preface

*Discontinuous Galerkin* (DG) methods are a class of *nonconforming* finite element approximations allowing piecewise regular solutions with possible discontinuities on the edges of the triangulation.

The first DG method was introduced in 1973 by Reed and Hill [6] to solve the hyperbolic equation modeling the neutron transportation. Since then, different DG methods have been designed to solve systems governed by partial differential equations (PDEs) of hyperbolic, elliptic, and/or parabolic type.

Due to their discontinuous character, the numerical solutions obtained by DG methods can handle very efficiently problems whose solutions present shocks. This is why they received a significant interest for first-order transport operators arising, for example, in fluid mechanics. The interested reader is oriented to the survey articles [11, 17] (and the references therein), where one can find the description and the main properties of the basic DG methods for second-order elliptic problems and conservation laws.

This book is devoted to carefully analyze the *propagation properties* of the numerical solutions generated by the DG space semi-discretization approximations of the $1-d$ linear *wave equation* and of its *Klein–Gordon* version. These properties are particularly important when dealing with applications motivated by control or inverse problems theory. In particular, we analyze the so-called observability property. It refers to the possibility to estimate the total energy of a given system by means of partial measurements on the solution taken by sensors placed on a subset of the space domain during a finite time interval. Of course, in order to make the observation mechanisms more efficient, in practice, this subset and the observation time have to be as small as possible. But when dealing with wave propagation phenomena, due to the *finite speed of propagation* of the energy of solutions, the observation region and time have to be large enough. A sharp necessary and sufficient *geometric control condition* (GCC) for the observation of the second-order linear wave equation in bounded smooth domains was given by Bardos–Lebeau–Rauch in [9]. This GCC requires, essentially, all bi-characteristic rays to reach the observability region during the observation period.

Once the observability is known to hold, by duality, using the so-called Hilbert uniqueness method (HUM) introduced by Lions (cf. [44]), the *exact controllability* of the system holds as well. *Controllability* refers to the possibility of steering the system to rest in finite time by means of controls located on the region where observations were made.

It is also well known that, unlike the *continuous wave* and *Klein–Gordon equations* for which the control/observability properties under consideration hold in a finite time, the simplest approximation methods on uniform meshes of these models (e.g., *finite differences* or *linear classical finite elements*) lead to undesired phenomena. Namely, the discrete analogues of the observability and controllability properties fail to be uniform with respect to the mesh size parameter. This is due to the fact that the discrete dynamics produces spurious solutions propagating with arbitrarily small velocities.

This is actually the main reason for the failure of the so-called discrete approach for controllability which is based on the natural idea that the continuous controls could be obtained by firstly controlling a convergent scheme for the wave or Klein–Gordon equations and then passing to the limit in the numerical control. But *the lack of uniformity of the propagation/observability properties makes this discrete approach to fail in general.*

The survey articles [27,28] provide a detailed presentation of the high-frequency pathologies for the finite difference approximations of the wave equation on uniform meshes and the needed *filtering mechanisms* aimed to reestablish the uniformity of the observability properties on suitable subspaces of initial data.

In this book, we analyze the *high-frequency pathologies* and propose several *filtering strategies* when the Laplace operator in the wave and Klein–Gordon equations is approximated by well-known *interior penalty* DG methods (cf. [6]).

As mentioned above, numerical solutions obtained by DG methods fail to be continuous at the mesh points, taking two different values on each side. In the limit as the mesh size goes to zero, the continuity of these discontinuous numerical solutions is *weakly enforced*, in the sense that the jumps on the edges of the triangulation are *penalized* by adding suitable *numerical fluxes*. In fact, the behavior of these discrete solutions is in practice better understood by taking *averages* and *jumps* of the discrete values at both sides of each node.

All along this book, we mainly focus on the *symmetric interior penalty discontinuous Galerkin* (SIPG) method. To facilitate the exposition, we restrict ourselves to the case of *linear approximations* on *uniform meshes*.

The content of this work is organized as follows. We begin by two introductory chapters, in which we present the continuous wave and Klein–Gordon equations, their finite difference, classical finite element and SIPG semi-discretizations under consideration, and the corresponding observability problems, followed by a small chapter containing bibliographical notes. In Chap. 4, we develop a careful Fourier analysis of the SIPG method which highlights the coexistence of two Fourier modes (*physical* and *spurious*) related to the two components of the numerical solution (*averages* and *jumps*). In Chap. 5, we rigorously construct high-frequency wave packets propagating arbitrarily slowly, localized on the critical points of each

Fourier mode. We also describe the effect of this lack of uniform propagation on the polynomial blowup (as the mesh size tends to zero) at any order of the observability constant. In Chap. 6, we construct efficient *filtering mechanisms*. In particular, we prove that the uniform observability property is recovered uniformly by considering initial data with *null jumps* and *averages given by a bi-grid filtering algorithm*. In Chap. 7, we explain how our results can be extended to other finite element methods, in particular to *quadratic finite elements*, *local DG* methods, and a version of the SIPG method adding *penalization on the normal derivatives* of the numerical solution at the grid points. We end up the book by Chap. 8, in which we mainly list some open problems related to the topics discussed in this monograph, and by Appendix A, in which we give the details of the technical proofs of some results stated in Chap. 4.

This book contributes to describe the state of the art on the propagation and control of the numerical methods for the wave-type equations and can be used as auxiliary material for courses on control of PDEs, on numerical methods for wave propagation, on Fourier analysis for numerical methods of PDEs, or on DG methods.

Graz, Styria, Austria — Aurora Marica
Bilbao, Basque Country, Spain — Enrique Zuazua
January 27, 2014

# Acknowledgments

Both authors were partially supported by the ERC Advanced Grant *FP7-246775 NUMERIWAVES*, Grant *MTM2011-29306* of MICINN Spain, Project *PI2010-04* of the Basque Government, and the *ESF Research Networking Programme OPTPDE*. Additionally, the work of the first author was supported by two grants of the Romanian Ministry of National Education (CNCS-UEFISCDI), i.e., projects *PN-II-ID-PCE-2012-4-0021 Variable Exponent Analysis: Partial Differential Equations and Calculus of Variations* and *PN-II-ID-PCE-2011-3-0075 Analysis, Control and Numerical Approximations of PDEs*, and by the *FWF- Fonds zur Förderung der Wissenschaftlichen Forschung* under SFB 32, *MOBIS - Mathematical Optimization and Applications in Biomedical Sciences* of the FWF—Austrian Science Fund. The last version of this work was completed while the second author was visiting the *Laboratoire Jacques-Louis Lions* with the support of the Paris City Hall within the Program *Research in Paris*.

# Contents

# Chapter 1
# Preliminaries

In this chapter, we present the *observability property* of the solutions of the Cauchy problem for the $1-d$ *wave* and *Klein–Gordon* equations and its relation with the *exact controllability problem*. Both properties are closely linked to the propagation properties of the solutions of these models. Thus, from a controllability point of view, the propagation properties of numerical solutions play also a key role.

One of the main contents of this book is a rigorous comparison of the *propagation properties* of the numerical solutions obtained by $P_1$-DG approximation methods of the $1-d$ wave equation with those that other more classical approximation methods like the *finite difference* (FD) or the *linear classical finite element* methods ($P_1$-FEM) yield. In this first chapter, we also present some well-known results in the literature stating the lack of uniform (with respect to the mesh size parameter) observability estimates for the FD and the $P_1$-FEM space semi-discrete versions of the wave equation and the efficiency of the corresponding filtering strategies based on the *Fourier truncation* and on the *bi-grid algorithm*.

## 1.1 Preliminaries on Control and Observation Problems

The notion of *observability* was introduced by the American-Hungarian engineer and mathematician *Rudolf Kalman* in the context of control for linear systems of ordinary differential equations (ODEs) (cf. [40]). It consists in estimating the total energy of the dynamical system under consideration by measuring the output signal over a finite time interval.

This notion is the dual of the *controllability* one that we present now.

Let us consider the system of nonhomogeneous ODEs:

$$\mathbf{Y}'(t) = A\mathbf{Y}(t) + B\mathbf{V}(t),\ t \in (0, T), \quad \mathbf{Y}(0) = \mathbf{Y}^0, \tag{1.1}$$

A. Marica and E. Zuazua, *Symmetric Discontinuous Galerkin Methods for 1 – D Waves*, SpringerBriefs in Mathematics, DOI 10.1007/978-1-4614-5811-1_1,

where $A$ is a $n \times n$-matrix, the solution (state) $\mathbf{Y}(t)$ is an $n$-dimensional vector, $B$ is the $n \times m$-control matrix, and $\mathbf{V}(t)$ is the $m$-dimensional control vector. The *exact controllability* (or, simply, *controllability*) problem consists in analyzing whether, for any initial data $\mathbf{Y}^0 \in \mathbb{R}^n$ and any final target $\mathbf{Y}^1 \in \mathbb{R}^n$, there exists a control vector $\mathbf{V}(t)$ so that the trajectory $\mathbf{Y}(t)$ of the system starting at $\mathbf{Y}^0$ can be driven to $\mathbf{Y}^1$ (i.e., $\mathbf{Y}(T) = \mathbf{Y}^1$) in time $T > 0$.

This exact controllability problem is equivalent to the *observability* one for the *adjoint system* that we introduce now.

Let us consider the *adjoint system* associated to (1.1) as

$$-\mathbf{U}'(t) = A^*\mathbf{U}(t),\ t \in (0,T), \quad \mathbf{U}(T) = \mathbf{U}^0. \tag{1.2}$$

Here, the superscript $*$ in $A^*$ denotes the transpose of the matrix $A$ and the unknown $\mathbf{U}(t)$ is an $n$-dimensional vector. Remark that the initial data in (1.2) is considered at the final time $T$, something that is irrelevant in this finite-dimensional setting, but that may be important when dealing with time-irreversible PDEs like the heat equation, for example.

The *observability property* for the system (1.2) can be characterized by the following *observability inequality*:

$$||\mathbf{U}(0)||^2_{\mathbb{R}^n} \le C(T)\int_0^T ||B^*\mathbf{U}(t)||^2_{\mathbb{R}^m}\,dt, \tag{1.3}$$

where $||\cdot||_{\mathbb{R}^n}$ is the Euclidean norm in $\mathbb{R}^n$ associated to the scalar product $(\cdot,\cdot)_{\mathbb{R}^n}$ defined by $(\mathbf{F},\mathbf{G})_{\mathbb{R}^n} = \sum_{j=1}^n F_j G_j$.

In practice, $m$, representing the number of measurements/controls in the observability/control problem, is taken smaller than $n$. The matrix $B^*$ is the so-called observability matrix.

The controllability and observability problems are equivalent. Both the exact controllability of (1.1) and the observability inequality (1.3) hold for any positive time $T > 0$ if and only if the following algebraic *Kalman rank condition* holds:

$$\text{rank}[B, AB, A^2B, \ldots, A^{n-1}B] = n. \tag{1.4}$$

When the system is controllable, for any initial and final data, there are many different controls. But there is only one of minimal $L^2(0,T)$-norm. This optimal control can be characterized by minimizing the functional

$$\mathscr{J}(\mathbf{U}^0) := \frac{1}{2}\int_0^T ||B^*\mathbf{U}(t)||^2_{\mathbb{R}^n} - (\mathbf{U}(0), \mathbf{Y}^0)_{\mathbb{R}^n}, \tag{1.5}$$

where $\mathbf{U}^0 \in \mathbb{R}^n$ is the datum at time $t = T$ in the uncontrolled (adjoint) problem (1.2) and $\mathbf{U}(t)$ is the corresponding solution. The optimal $L^2(0,T)$-control is then given by $\mathbf{V}(t) = \tilde{\mathbf{V}}(t) := B^*\tilde{\mathbf{U}}(t)$, where $\tilde{\mathbf{U}}(t)$ is the solution of (1.2) corresponding to the minimum $\tilde{\mathbf{U}}^0$ of $\mathscr{J}$.

Note that the existence and uniqueness of a minimizer for the quadratic functional $\mathscr{J}$ is guaranteed by the *direct method in the calculus of variations* (cf. [21], Theorem 3.30, pp. 106), requiring *continuity*, *convexity*, and *coercivity* of $\mathscr{J}$. The observability inequality ensures, precisely, the *coercivity* requirement.

The fact that the optimal control of a system can be constructed explicitly in terms of the minimizer of a functional depending on the solution of the corresponding uncontrolled problem and the initial data to be controlled is the essence of the *principle of duality* in the control of ODEs (cf. [40]) and of the *Hilbert uniqueness method* (HUM) in the theory of PDEs (cf. [44]). For more details concerning this duality argument, the interested reader is referred to the survey articles [27, 28, 58, 70].

In practice, this duality strategy has many applications (see, e.g., [16] or [64]). In engineering, for instance, control systems incorporate both *actuators* and *sensors*. *Actuators* transform input signals into particular types of motions (e.g., linear or circular ones). Concrete examples of such devices are electrical motors, pneumatic or piezoelectric actuators, hydraulic pistons, relays, comb drives, etc. *Sensors* or *detectors* measure physical quantities and convert them into signals that can be read by *observers* or *instruments*. An efficient device should place strategically the sensors, so to detect any possible signal and to regulate properly the future motion of the device. The optimal placement of the sensors on some components of the system is aimed to make it efficient at the minimal cost and, in the simple situation of the ODE systems (1.1) and (1.2), this is possible to be achieved whenever the rank condition (1.4) is fulfilled

While a linear finite system of ODEs is controllable/observable in any positive time $T$ if and only if the Kalman condition (1.4) holds (since (1.4) does not depend on $T$, but only on the matrices $A$ and $B$ involved in the system), infinite-dimensional systems of hyperbolic PDEs can require a finite but sufficiently large time to be controllable/observable. This happens due to the finite speed of propagation for the solutions of these hyperbolic systems, so that information needs a minimal time to travel from a certain part of the domain to the control/observation region. *The observability/controllability time is the one needed by the support of all possible solutions to enter the observability/controllability region.*

This is precisely the Hamiltonian point of view giving rise to the *geometric control condition* (GCC), coming from classical mechanics and requiring all *bi-characteristic rays to enter the observability region* (cf. [9]).

To be more precise, consider the nonhomogeneous problem associated to the linear $1-d$ wave equation on the interval $(0,1)$:

$$\begin{cases} y_{tt} - y_{xx} = 0, & x \in (0,1),\ t \in (0,T), \\ y(0,t) = 0,\ y(1,t) = v(t), & t \in (0,T), \\ y(x,0) = y^0(x), \quad y_t(x,0) = y^1(x), & x \in (0,1). \end{cases} \tag{1.6}$$

The left endpoint of the string is fixed to have null value, while its right endpoint is controlled by means of the control function $v \in L^2(0,T)$. If, additionally to $v \in L^2(0,T)$, the initial data $(y^0, y^1)$ in (1.6) belong to $L^2 \times H^{-1}(0,1)$, then it is well known (cf. [44]) that $y \in C([0,T]; L^2(0,1)) \cap C^1([0,T]; H^{-1}(0,1))$.

Given arbitrary initial data $(y^0, y^1) \in L^2 \times H^{-1}(0,1)$, the *null-controllability problem* in time $T$ consists in finding a control $v \in L^2(0,T)$ so that, at time $T$, the solution $y$ of (1.6) verifies the equilibrium condition

$$y(x,T) = y_t(x,T) = 0 \text{ for all } x \in (0,1). \tag{1.7}$$

Due to the *reversibility in time* of the wave equation, the above-stated *null-controllability* of (1.6) is equivalent to its *exact controllability*. This last property means that, given any $(\tilde{y}^0, \tilde{y}^1) \in L^2 \times H^{-1}(0,1)$, the control $v$ has to be chosen such that the solution of (1.6) satisfies $y(x,T) = \tilde{y}^0(x)$ and $y_t(x,T) = \tilde{y}^1(x)$ instead of the equilibrium condition corresponding to *null-controllability*.

The optimal $L^2(0,T)$-null control is constructed by minimizing the functional

$$\mathscr{J}(u^0,u^1) := \frac{1}{2}\int_0^T |u_x(1,t)|^2\,dt + \int_0^1 y^0(x)u_t(x,0)\,dx - \langle y^1, u(\cdot,0)\rangle_{H^{-1}\times H_0^1}, \tag{1.8}$$

where $\langle\cdot,\cdot\rangle_{H^{-1}\times H_0^1}$ is the *duality pairing* between $H^{-1}(0,1)$ and $H_0^1(0,1)$ and $u$ is the solution of the following adjoint wave equation corresponding to the data $(u^0,u^1) \in H_0^1 \times L^2(0,1)$ at the final time $T$:

$$\begin{cases} u_{tt} - u_{xx} = 0, & x \in (0,1),\ t \in (0,T), \\ u(0,t) = u(1,t) = 0, & t \in (0,T) \\ u(x,T) = u^0(x), \quad u_t(x,T) = u^1(x), & x \in (0,1). \end{cases} \tag{1.9}$$

The optimal control is $v(t) = \tilde{v}(t) := \tilde{u}_x(1,t)$, where $\tilde{u}$ is the solution of (1.9) corresponding to the minimizer $(\tilde{u}^0, \tilde{u}^1) \in H_0^1 \times L^2(0,1)$ of $\mathscr{J}$ in (1.8). Here, by *optimal* we refer to the control of minimal $L^2(0,T)$-norm fulfilling the equilibrium requirement (1.7) at $t = T$.

The *coercivity property* guaranteeing the existence of a unique minimizer of $\mathscr{J}$ in (1.8) is ensured when the observability inequality below holds for all $(u^0,u^1) \in H_0^1 \times L^2(0,1)$:

$$\mathscr{E}(u^0,u^1) \le C(T)\int_0^T |u_x(1,t)|^2\,dt, \tag{1.10}$$

where

$$\mathscr{E}(u^0,u^1) := \frac{1}{2}\left(||u^0||^2_{H_0^1} + ||u^1||^2_{L^2}\right) \tag{1.11}$$

is the time conservative *total energy* of the solution $u(x,t)$ of (1.9) corresponding to the initial data $(u^0,u^1) \in H_0^1 \times L^2(0,1)$.

It is well known [9, 44] that (1.10) holds for all time $T \ge T^\star := 2$. The lower bound on the observability/controllability time $T^\star = 2$ is sharp.

From a numerical analysis point of view, the most classical finite difference approximation schemes of the wave equation are systems of ODEs of the form (1.2) (or (1.1), in the controlled version) and the following natural questions arise: *Are they controllable/observable? Do the corresponding controls converge as the dimension of approximation tends to infinity (while the mesh size tends to zero) to the control of the wave equation?* As we shall see in the next section, the answer to the first question is affirmative, while for the second one is of negative nature. It is for that reason that the classical numerical approximation schemes have to be modified (essentially by filtering out the high-frequency components), so to ensure the convergence of the discrete controls.

## 1.2 Classical Approximation Schemes: FD and $P_1$-FEM

Consider the semi-discrete *finite difference* (FD) and *piecewise linear and continuous finite element* ($P_1$-FEM) approximation schemes (continuous in time and discrete in space) of the controlled/adjoint wave equations (1.6) and (1.9). They take the form

$$M_s^h \mathbf{y}_{tt}^h(t) + R^h \mathbf{y}^h(t) = \mathbf{f}^h(t), \quad \mathbf{y}^h(0) = \mathbf{y}^{h,0}, \ \mathbf{y}_t^h(0) = \mathbf{y}^{h,1}, \tag{1.12}$$

and, respectively,

$$M_s^h \mathbf{u}_{tt}^h(t) + R^h \mathbf{u}^h(t) = 0, \quad \mathbf{u}^h(T) = \mathbf{u}^{h,0}, \ \mathbf{u}_t^h(T) = \mathbf{u}^{h,1}. \tag{1.13}$$

The subscript $s$ takes the value $s = 1$ for the FD scheme and $s = \infty$ for the $P_1$-FEM one. The reason to associate $s = 1$ and $s = \infty$ to the FD and the $P_1$-FEM approximations will be made clear after the analysis of the discontinuous Galerkin approximation methods and after its comparison to the FD and $P_1$-FEM approximations will be carried out in Chap. 2 of this book. Briefly, this is so since the FD and the $P_1$-FEM methods can be viewed as limit cases (as the penalty parameter $s$ tends to one or to infinity) of a whole family of DG methods.

In systems (1.12) and (1.13), $h = 1/(N+1)$ is the mesh size, $N$ is the number of internal grid points, $\mathbf{y}^h(t)$ and $\mathbf{u}^h(t)$ are $N$-dimensional vectors of unknowns, $R^h$ is the *stiffness matrix* of the three-point centered scheme approximating the Laplacian (tridiagonal, of dimension $N$, taking value $2/h^2$ on the main diagonal and $-1/h^2$ on the upper and lower diagonals), $M_s^h$ is the *mass matrix* (i.e., the identity matrix for $s = 1$ corresponding to the FD approximation and the tridiagonal matrix of dimension $N$, having $2/3$ on the main diagonal and $1/6$ on the lower and upper diagonals for $s = \infty$, corresponding to the $P_1$-FEM approximation), and $\mathbf{f}^h(t) := (0, \ldots, 0, v^h(t)/h^2)$ is the $N$-dimensional vector containing the boundary data $y_{N+1}(t) = v^h(t)$.

By setting $\mathbf{Y}(t) = \mathbf{Y}^h(t) := (\mathbf{y}^h(t), \mathbf{y}_t^h(t))$ and $\mathbf{U}(t) = \mathbf{U}^h(t) := (\mathbf{u}^h(t), \mathbf{u}_t^h(t))$, the semi-discrete wave equations (1.12) and (1.13) can be transformed into first-order systems in time of the forms (1.1) and (1.2), in which

$$A = A^h := \begin{pmatrix} 0_{N,N} & I_N \\ Q_s^h & 0_{N,N} \end{pmatrix}, \quad B = B^h := \begin{pmatrix} 0_{N,1} \\ (M_s^h)^{-1}q \end{pmatrix}, \quad q := \begin{pmatrix} 0_{N-1,1} \\ 1/h^2 \end{pmatrix},$$

where $Q_s^h := -(M_s^h)^{-1}R^h$, $0_{m,n}$ is the zero $m \times n$-dimensional matrix and $I_N$ is the $N$-dimensional identity matrix.

The Kalman rank condition (1.4) is verified, so that one can guarantee observability in any finite time $T$. More precisely, the observability inequality corresponding to system (1.13) is as follows:

$$\mathscr{E}_s^h(\mathbf{u}^{h,0}, \mathbf{u}^{h,1}) \leq C_s^h(T) \int_0^T \left| \frac{u_N(t)}{h} \right|^2 dt. \tag{1.14}$$

Here, $\mathscr{E}_s^h(\mathbf{u}^h(t), \mathbf{u}_t^h(t))$ given below is the (time conservative) total energy of the solution of (1.13):

$$\mathscr{E}_s^h(\mathbf{u}^h(t), \mathbf{u}_t^h(t)) := \frac{1}{2}(M_s^h \mathbf{u}_t^h(t), \mathbf{u}_t^h(t)) + \frac{1}{2}(R^h \mathbf{u}^h(t), \mathbf{u}^h(t)).$$

Let us now give a short proof of the fact that inequality (1.14) holds for all $T > 0$ and $h > 0$ for both $s = 1$ and $s = \infty$ based on the analysis of the Kalman rank condition (1.4).

Indeed, let $K := [B, AB, \ldots, A^{2N-1}B]$ be the *Kalman matrix*. It is enough to prove that $\det(K) \neq 0$. It is not difficult to see that

$$A^{2k} = \begin{pmatrix} (Q_s^h)^k & 0_{N,N} \\ 0_{N,N} & (Q_s^h)^k \end{pmatrix} \text{ and } A^{2k+1} = \begin{pmatrix} 0_{N,N} & (Q_s^h)^k \\ (Q_s^h)^{k+1} & 0_{N,N} \end{pmatrix}$$

for all $0 \leq k \leq N-1$, so that $K = [K_1, K_2, \ldots, K_{2N}]$, with

$$K_{2j-1} := \begin{pmatrix} 0_{N,1} \\ (Q_s^h)^{j-1}(M_s^h)^{-1}q \end{pmatrix} \text{ and } K_{2j} := \begin{pmatrix} (Q_s^h)^{j-1}(M_s^h)^{-1}q \\ 0_{N,1} \end{pmatrix}, \quad 1 \leq j \leq N.$$

Using Laplace's theorem (cf. [60], pp. 57) to develop $\det(K)$ over the even columns and the last $N$ rows, we get $\det(K) = (-1)^{N^2+N(N+1)/2}(\det(\tilde{K}))^2$, where

$$\tilde{K} := [(M_s^h)^{-1}q, Q_s^h(M_s^h)^{-1}q, \ldots, (Q_s^h)^{N-1}(M_s^h)^{-1}q].$$

Remark that the mass and stiffness matrices admit the spectral decomposition $M_s^h := ED_M E^{-1}$ and $R^h := ED_R E^{-1}$, where $E = (e_{j,k})_{1 \leq k,j \leq N}$ ($e_{j,k} := \sqrt{2h}\sin(k\pi x_j)$) is the unitary matrix of eigenvectors (i.e., $E$ is invertible,

$E^{-1} = E^*$, and $\det(E) = 1$) and $D_M$ and $D_R$ are diagonal matrices containing the eigenvalues $(\Lambda_M^1, \dots, \Lambda_M^N)$ and $(\Lambda_R^1, \dots, \Lambda_R^N)$ of $M_s^h$ and $R^h$ ($\Lambda_M^k := 1$ for $s = 1$ and $\Lambda_M^k := (2 + \cos(k\pi h))/3$ for $s = \infty$, $\Lambda_R^k := 4\sin^2(k\pi h/2)/h^2$)). Set $\Lambda^k := \Lambda_R^k/\Lambda_M^k$. It is not difficult to show that

$$\begin{aligned}\det(\tilde{K}) &= \det\left(E\left((-D_M^{-1}D_R)^{j-1}D_M^{-1}E^*q\right)_{1\le j\le N}\right)\\ &= \det\left(\left(\frac{(-\Lambda^j)^{k-1}e_{j,N}}{h^2\Lambda_M^j}\right)_{1\le j,k\le N}\right)\\ &= \frac{e_{1,N}\cdots e_{N,N}}{h^{2N}\Lambda_M^1\cdots\Lambda_M^N}(-1)^{N(N-1)/2}\prod_{1\le j<k\le N}(\Lambda^k-\Lambda^j),\end{aligned}$$

so that $\det(\tilde{K}) \neq 0$ and $\operatorname{rank}(K) = 2N$.

Another possible proof of the fact that $\operatorname{rank}(K) = 2N$, valid only for the finite difference scheme, is as follows. In fact, it is sufficient to show directly that (1.14) holds. Since we are in a finite-dimensional setting, it is enough to show that the right-hand side term of (1.14) is a norm. To do so, it is sufficient to prove that if $u_N(t) = 0$ for all $t \in [0, T]$, then $\mathbf{u}^h(t) = 0$. But, using the FD scheme (1.13) at the node $j = N$ and the boundary condition at the node $j = N + 1$, it is easy to prove that this implies that $u_{N-1}(t) = 0$ for all $t \in [0, T]$. Iterating this argument, it follows that $u_j(t) = 0$ for all $t \in [0, T]$ and $0 \le j \le N + 1$.

However, the constant $C_s^h(T)$ in (1.14) depends on $h$, and, even more, it blows up as $h \to 0$ for any finite $T > 0$ (cf. [27] and the references therein). This happens, for example, when the initial data $\mathbf{U}^0$ is concentrated on Fourier modes corresponding to the highest eigenvalues, for which the velocity of propagation is not one, as in the continuous case, but rather very close to zero (we refer to the constructions of highly oscillatory Gaussian wave packets in [28,49], proving the blowup of $C(T)$ in (1.14) with, at least, an arbitrary large polynomial order for both $s = 1$ and $s = \infty$, or the estimates on bi-orthogonal sequences in [56], showing the exponential divergence of $C_s^h(T)$ for $s = 1$ and for all values of $T$).

As a consequence of these high-frequency pathologies, the following quadratic functional fails to be uniformly coercive as the mesh size parameter $h$ goes to zero:

$$\mathcal{J}^h(\mathbf{u}^{h,0}, \mathbf{u}^{h,1}) := \frac{1}{2}\int_0^T \left|\frac{u_N(t)}{h}\right|^2 dt + (\mathbf{y}^{h,1}, \mathbf{u}^h(0)) - (\mathbf{y}^{h,0}, \mathbf{u}_t^h(0)).$$

Here, $(\mathbf{y}^{h,0}, \mathbf{y}^{h,1})$ are the data to be controlled to rest in (1.12), $(\mathbf{u}^{h,0}, \mathbf{u}^{h,1})$ is the final data in (1.13), and $\mathbf{u}^h(t)$ is the corresponding solution. The coercivity constant of $\mathcal{J}^h$ is $1/(4C^h(T))$, where $C^h(T)$ is the observability constant in (1.14). Thus, when dealing with the whole class of solutions $\mathbf{u}^h(t)$ of (1.13), this coercivity constant tends to zero exponentially as $h \to 0$.

Accordingly, the corresponding optimal controls $\tilde{\mathbf{v}}^h(t)$ in (1.12) obtained by minimizing the functional $\mathcal{J}^h$ diverge as $h$ goes to zero. This blowup is motivated by the fact that the controls have to handle not only the components of numerical solutions which are very close to the ones of the continuous problem but also the spurious ones, appearing only at the discrete level and oscillating at wavelengths of order $h$.

This pathological behavior can be explained by comparing the *dispersion relations* of the continuous and of the discrete models. Thus, for any $s \in \{1, \infty\}$, one can write the solution of the discrete system (1.13) as

$$\mathbf{u}^h(t) = \sum_{\pm} \sum_{k=1}^{N} \frac{1}{2} \left( \hat{u}^{k,0} \pm \frac{\hat{u}^{k,1}}{i \hat{\lambda}_s^{h,k}} \right) \exp(it\hat{\lambda}_s^{h,k}) \hat{\boldsymbol{\varphi}}^{h,k}, \tag{1.15}$$

where $\hat{\lambda}_s^{h,k} := \sqrt{\hat{\Lambda}_s^{h,k}}$ and $(\hat{\Lambda}_s^{h,k}, \hat{\boldsymbol{\varphi}}^{h,k})$, $1 \leq k \leq N$, are the solutions of the following *spectral problem*:

$$R^h \hat{\boldsymbol{\varphi}}^{h,k} = \hat{\Lambda}_s^{h,k} M_s^h \hat{\boldsymbol{\varphi}}^{h,k}. \tag{1.16}$$

More explicitly, $\hat{\Lambda}_s^{h,k} = \hat{\Lambda}_{s,ph}^h(k\pi)$ and $\hat{\boldsymbol{\varphi}}^{h,k} = (\sqrt{2}\sin(k\pi x_j))_{1 \leq j \leq N}$, where

$$\hat{\Lambda}_{1,ph}^h(\xi) := \frac{4}{h^2} \sin^2\left(\frac{\xi h}{2}\right) \text{ and } \hat{\Lambda}_{\infty,ph}^h(\xi) := \frac{4}{h^2} \sin^2\left(\frac{\xi h}{2}\right) \frac{3}{2 + \cos(\xi h)}. \tag{1.17}$$

Although for both FD and $P_1$-FEM approximations the subscript $ph$ has no relevance, we inherit this notation from the discontinuous Galerkin approximations, for which there are two Fourier modes. One of them is called the *physical mode*, denoted $\hat{\Lambda}_{s,ph}^h(\xi)$, and coincides with $\hat{\Lambda}_{1,ph}^h(\xi)$ and $\hat{\Lambda}_{\infty,ph}^h(\xi)$ in the limit as $s \to 1$ or $s \to \infty$.

One of the main *differences* between the discrete diagrams $\hat{\lambda}_{s,ph}^h(\xi) := \sqrt{\hat{\Lambda}_{s,ph}^h(\xi)}$, $s \in \{1, \infty\}$ and the continuous one $\hat{\lambda}(\xi) := \xi$ is related to the *group velocity*. For the continuous case, $\partial_\xi \hat{\lambda}(\xi) \equiv 1$ for all $\xi \in \mathbb{R}$ and this agrees with the fact that continuous waves propagate only along characteristic lines of unit slope, $x(t) = x^* \pm t$. However, the discrete group velocity $\partial_\xi \hat{\lambda}_{s,ph}^h(\xi)$ varies continuously from $\xi = 0$ to $\xi = \pi/h$. When $s = 1$, it decreases from 1 (at $\xi = 0$) to 0 (at $\xi = \pi/h$), while for $s = \infty$, the group velocity first increases from 1 (at $\xi = 0$) to $\sqrt{2}$ ($\xi = 2\pi/(3h)$) and, after that, decreases from $\sqrt{2}$ (at $\xi = 2\pi/(3h)$) to 0 (at $\xi = \pi/h$). See Fig. 1.1 for the case $h = 1$. The fact that the group velocity can vanish means that, for the discrete system (1.13), there are wave packets propagating along rays of slope arbitrarily close to zero, spending an arbitrarily large time to cross the domain $(0, 1)$ and making then impossible the observability property (1.14) to hold uniformly as $h \to 0$ in a finite time $T$.

Several *filtering mechanisms* aimed to reestablish the uniformity of the propagation properties have been proposed in the literature: the *Fourier truncation method*

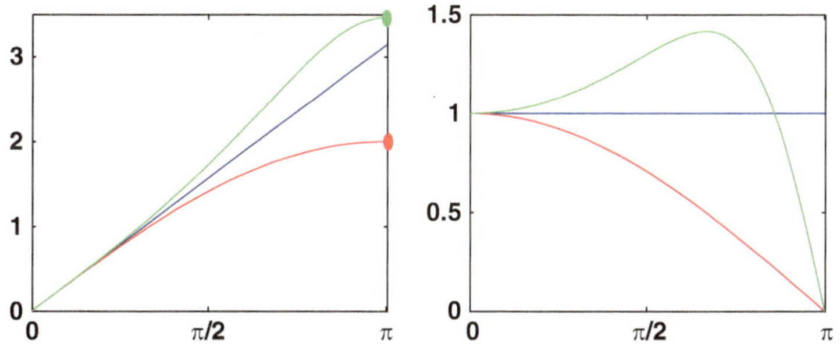


**Fig. 1.1** *Left*: in *blue/red/green*, the continuous dispersion relation $\hat{\lambda}(\xi) = \xi$ and the ones corresponding to the FD and $P_1$-FEM schemes, $\hat{\lambda}^1_{1,ph}(\xi)$ and $\hat{\lambda}^1_{\infty,ph}(\xi)$. At the *marked points*, the corresponding group velocity $\partial_\xi \hat{\lambda}^1_{s,ph}(\xi)$ vanishes. *Right*: the three group velocities

(cf. [39]), the *bi-grid algorithms* [31, 33, 37, 45, 59], the *numerical viscosity method* [57, 66], the *Tychonoff regularization* [31, 33], and the *mixed-type finite element methods* (cf. [15]). The *Fourier truncation method* reduces the class of solutions of (1.13) so that they involve only Fourier modes of indices $k \leq \lfloor \delta/h \rfloor$ in (1.15) for some $\delta \in (0,1)$ (here, $\lfloor r \rfloor$ is the integer part of the real number $r$). Then, as proved in [39], for all $\delta \in (0,1)$, all $T \geq T^\star_{s,\delta} := 2/\min\{1, \partial_\xi \hat{\lambda}^1_{s,ph}(\pi\delta)\}$, and all solutions $\mathbf{u}^h(t)$ of (1.13) corresponding to initial data in this truncated class, the observability property (1.14) holds uniformly as $h \to 0$. The *bi-grid algorithm* produces similar results in the subclass of slowly oscillating initial data obtained by *linear interpolation* from a *twice coarser grid* for all $T > T^\star_{s,1/2}$ (cf. [59]).

In this book, we are interested in the observability/controllability problem for more sophisticated numerical approximation methods of the wave equation and, in particular, the so-called *discontinuous Galerkin (DG) finite element* methods. But before introducing these numerical approximation schemes and their main features, let us present rigorously the continuous models: the *wave* and *Klein–Gordon* equations on the whole real line.

## 1.3 The Continuous Wave Equation

Consider the following Cauchy problem for the $1-d$ *wave equation*:

$$\begin{cases} u_{tt}(x,t) - u_{xx}(x,t) = 0, & x \in \mathbb{R},\ t > 0 \\ u(x,0) = u^0(x),\ u_t(x,0) = u^1(x), & x \in \mathbb{R}. \end{cases} \tag{1.18}$$

This problem is well posed in $\dot{H}^1(\mathbb{R}) \times L^2(\mathbb{R})$ so that, for all $(u^0, u^1) \in \dot{H}^1(\mathbb{R}) \times L^2(\mathbb{R})$, there exists a unique solution $u \in C([0,\infty), \dot{H}^1(\mathbb{R})) \cap C^1([0,\infty), L^2(\mathbb{R}))$.

Here, $\dot{H}^1(\mathbb{R})$ denotes the *homogeneous Sobolev space* obtained as the completion of $C_c^\infty(\mathbb{R})$ with respect to the semi-norm $\|\cdot\|_{\dot{H}^1(\mathbb{R})} := \|\partial_x \cdot\|_{L^2(\mathbb{R})}$ (cf. [8]). The *total energy* of the solution given below is conserved in time:

$$\mathscr{E}^W(u(\cdot,t), u_t(\cdot,t)) := \frac{1}{2}\big(\|u(\cdot,t)\|^2_{\dot{H}^1(\mathbb{R})} + \|\partial_t u(\cdot,t)\|^2_{L^2(\mathbb{R})}\big). \tag{1.19}$$

The *observability problem* consists in determining whether the *total energy of solutions* can be estimated in terms of the *energy concentrated on some subset of the spatial domain* where waves propagate, the so-called observation region.

For the continuous problem (1.18), it is well known (cf. [68]) that the observability property holds when the observation region is the complement of a compact set. More precisely, we can obtain the following observability result:

**Theorem 1.1.** *Consider the observation subset* $\Omega := \mathbb{R} \setminus (-1,1)$*. For any observability time* $T > T^\star := 2$ *and for all solutions* $u = u(x,t)$ *of (1.18) with initial data* $(u^0, u^1) \in \dot{H}^1(0,1) \times L^2(0,1)$*, there exists a constant* $C(T) > 0$ *such that the following observability inequality holds:*

$$\mathscr{E}^W(u^0, u^1) \leqslant C(T) \int_0^T \mathscr{E}^W_\Omega(u(\cdot,t), u_t(\cdot,t))\, dt. \tag{1.20}$$

*Here,* $\mathscr{E}^W_\Omega(f^0, f^1)$ *is the energy concentrated in* $\Omega$ *given below:*

$$\mathscr{E}^W_\Omega(f^0, f^1) = \frac{1}{2}\int_\Omega (|f^0_x(x)|^2 + |f^1(x)|^2)\, dx.$$

*Proof.* The simplest way to prove the observability inequality (1.20) is using the classical *d'Alembert formula* for the solution $u$ of (1.18):

$$u(x,t) = \frac{1}{2}(u^0(x+t) + u^0(x-t)) + \frac{1}{2}\int_{x-t}^{x+t} u^1(y)\, dy. \tag{1.21}$$

Therefore, for all $x \in \mathbb{R}$ and $t \in [0,T]$, we get

$$\frac{1}{2}(|u_t(x,t)|^2 + |u_x(x,t)|^2) = \frac{1}{4}\big|(u^0)'(x+t) + u^1(x+t)\big|^2 + \frac{1}{4}\big|(u^0)'(x-t) - u^1(x-t)\big|^2. \tag{1.22}$$

By decomposing the total energy $\mathscr{E}^W(u(\cdot,t), u_t(\cdot,t))$ as

$$\mathscr{E}^W(u(\cdot,t), u_t(\cdot,t)) = \mathscr{E}^W_\Omega(u(\cdot,t), u_t(\cdot,t)) + \mathscr{E}^W_I(u(\cdot,t), u_t(\cdot,t)),$$

with $I = [-1,1]$, and taking into account the time conservation property of the total energy, we obtain the identity

$$\int_0^T \mathscr{E}^W_\Omega(u(\cdot,t), u_t(\cdot,t))\, dt = T\mathscr{E}^W(u^0, u^1) - \int_0^T \mathscr{E}^W_I(u(\cdot,t), u_t(\cdot,t))\, dt.$$

Following (1.22), the energy localized on $I = [-1, 1]$ can be written as

$$\begin{aligned}
&\int_0^T \mathscr{E}_I^W(u(\cdot,t), u_t(\cdot,t))\,dt \\
&\quad = \frac{1}{4}\int_0^T \int_{-1+t}^{1+t} \left|(u^0)'(y) + u^1(y)\right|^2 dy\,dt \\
&\quad + \frac{1}{4}\int_0^T \int_{-1-t}^{1-t} \left|(u^0)'(y) - u^1(y)\right|^2 dy\,dt.
\end{aligned}$$

By changing the order of integration, we obtain

$$\begin{aligned}
\int_0^T \int_{-1+t}^{1+t} \left|(u^0)'(y) + u^1(y)\right|^2 dy\,dt &= \int_{-1}^{1} (y+1)\left|(u^0)'(y) + u^1(y)\right|^2 dy \\
&+ 2\int_1^{T-1} \left|(u^0)'(y) + u^1(y)\right|^2 dy \\
&+ \int_{T-1}^{T+1} (T+1-y)\left|(u^0)'(y) + u^1(y)\right|^2 dy \\
&\le 2\int_{-1}^{T+1} \left|(u^0)'(y) + u^1(y)\right|^2 dy
\end{aligned}$$

and similarly

$$\int_0^T \int_{-1-t}^{1-t} \left|(u^0)'(y) - u^1(y)\right|^2 dy\,dt \le 2\int_{-1-T}^{1} \left|(u^0)'(y) - u^1(y)\right|^2 dy.$$

Consequently,

$$\begin{aligned}
\int_0^T \mathscr{E}_I^W(u(\cdot,t), u_t(\cdot,t))\,dt &\le \frac{1}{2}\int_{-1-T}^{1+T} \left(\left|(u^0)'(y)+u^1(y)\right|^2+\left|(u^0)'(y)-u^1(y)\right|^2\right) dy \\
&= \int_{-1-T}^{1+T} \left(\left|(u^0)'(y)\right|^2+\left|u^1(y)\right|^2\right) dy \le 2\mathscr{E}^W(u^0, u^1).
\end{aligned}$$ □

Similar results to (1.20) hold for any $1-d$ exterior domain $\Omega := \mathbb{R} \setminus (a,b)$, where $(a,b)$ is any finite interval, provided that $T > T^\star := b-a$, and also in several space dimensions. The best constant $C(T)$ in (1.20) is referred to as the *observability constant*.

In view of the finite velocity of propagation (=1, in this case), the *characteristic time* $T^\star := 2$ needed for (1.20) to hold is sharp. More precisely, $T^\star = 2$ is the time needed by a wave packet supported in an arbitrarily narrow neighborhood of one of the endpoints of the interval $(-1,1)$ at time $t = 0$ to reach the other endpoint, traveling along the characteristic lines $x(t) = x \pm t$.

When $T < T^\star := 2$, one can use the d'Alembert formula (1.21) to show that (1.20) fails because of the existence of nontrivial solutions vanishing in $\Omega$ for all $0 < t < T$.

The *observability property* (1.20) is motivated in particular by *controllability problems*. More precisely, by means of the HUM introduced by Lions in [44], the observability inequality (1.20) for the system (1.18) is equivalent to the following *controllability property*:

**Theorem 1.2.** *For all $T > T^\star := 2$ and all $(y^0, y^1) \in \dot{H}^1(\mathbb{R}) \times L^2(\mathbb{R})$, there exists a control function $c \in L^2(\Omega \times (0,T))$ such that the solution of the following nonhomogeneous problem*

$$\begin{cases} y_{tt}(x,t) - y_{xx}(x,t) = c(x,t)\chi_\Omega(x,t), & x \in \mathbb{R},\ t \in (0,T] \\ y(x,0) = y^0(x),\ y_t(x,0) = y^1(x), & x \in \mathbb{R}, \end{cases} \tag{1.23}$$

*satisfies the equilibrium condition at time $T$, i.e.,*

$$y(x,T) = y_t(x,T) = 0 \textit{ for all } x \in \mathbb{R}. \tag{1.24}$$

The optimal control $c = \tilde{c} \in L^2(\Omega \times (0,T))$, the one of minimal $L^2(\Omega \times (0,T))$-norm, takes the form

$$\tilde{c}(x,t) = \tilde{u}(x,t),$$

where $\tilde{u}(x,t)$ is the solution of (1.18) corresponding to the minimizer $(\tilde{u}^0, \tilde{u}^1) \in L^2(\mathbb{R}) \times \dot{H}^{-1}(\mathbb{R})$ of the quadratic functional

$$\begin{aligned} \mathscr{J}(u^0, u^1) &= \frac{1}{2}\int_0^T \int_\Omega |u(x,t)|^2\,dx\,dt \\ &\quad + \int_{\mathbb{R}} y_t(x,0)u(x,0)\,dx - \int_{\mathbb{R}} u_t(x,0)y(x,0)\,dx. \end{aligned}$$

These issues concerning the continuous wave equation are by now well understood and have been the object of intensive research. We refer to [32] or to [71] for a recent survey on this and many other closely related topics.

The observability inequality (1.20) corresponds to a different functional setting for controllability. Indeed, according to it, the control belongs to a larger class of, roughly, $H^{-1}$ controls, while the controlled data lie in $L^2(\mathbb{R}) \times \dot{H}^{-1}(\mathbb{R})$. Such a controllability statement can be proved directly out of the observability inequality (1.20) by means of the minimization principle above but conveniently adjusting the functional $\mathscr{J}$ to that functional setting. Once this is done, using the general methodology in [26] to get smoother controls for smoother data for abstract conservative systems, one gets the controllability result in the more convenient functional setting of $L^2$-controls and finite energy solutions. By duality, this leads to an observability inequality in $L^2(\Omega \times (0,T))$ of the form

$$||(\tilde{u}^0, \tilde{u}^1)||^2_{L^2(\mathbb{R})\times\dot{H}^{-1}(\mathbb{R})} \le C \int_0^T \int_\Omega |u|^2 dx dt. \tag{1.25}$$

## 1.4 The Continuous Klein–Gordon Equation

Consider the Cauchy problem for the $1-d$ *Klein–Gordon equation*:

$$\begin{cases} u_{tt}(x,t) - u_{xx}(x,t) + u(x,t) = 0, \; x \in \mathbb{R}, \; t > 0 \\ u(x,0) = u^0(x), \; u_t(x,0) = u^1(x), \; x \in \mathbb{R}, \end{cases} \tag{1.26}$$

which is well posed in $H^1(\mathbb{R}) \times L^2(\mathbb{R})$. The following *total energy* of the solution is also time conservative:

$$\mathscr{E}^K(u(\cdot,t), u_t(\cdot,t)) := \frac{1}{2}\big(\|u(\cdot,t)\|^2_{H^1(\mathbb{R})} + \|\partial_t u(\cdot,t)\|^2_{L^2(\mathbb{R})}\big). \tag{1.27}$$

For the continuous problem (1.26), we consider the same observation subset as for the wave equation (1.18), i.e., $\Omega := \mathbb{R} \setminus (-1,1)$. The following observability result holds:

**Theorem 1.3.** *For any $T > T^\star := 2$ there exists a constant $C(T) > 0$ such that the following observability inequality holds for all solutions $u = u(x,t)$ of the Klein–Gordon equation (1.26) with initial data $(u^0, u^1) \in L^2(0,1) \times H^{-1}(0,1)$:*

$$||(u^0, u^1)||^2_{L^2(\mathbb{R})\times H^{-1}(\mathbb{R})} \leqslant C(T) \int_0^T \int_\Omega |u|^2 \, dx \, dt.$$

*Proof.* This observability inequality for the Klein–Gordon system can be obtained out of the observably inequality (1.25) for the wave equation. It suffices to write the solution of the Klein–Gordon equation as a perturbation of the wave one, by means of the variations of constant formula. This yields the desired inequality with an extra reminder of order $H^{-1}$ in the complement of the observed region. This extra term can be absorbed by a classical compactness–uniqueness argument (cf. [44]). □

This observability inequality leads to the following *controllability property*:

**Theorem 1.4.** *For all $T > T^\star = 2$ and all $(y^0, y^1) \in H^1(\mathbb{R}) \times L^2(\mathbb{R})$, there exists a control function $c \in L^\infty(0,T; L^2(\Omega))$ such that the solution of the following nonhomogeneous problem:*

$$\begin{cases} y_{tt}(x,t) - y_{xx}(x,t) + y(x,t) = c(x,t)\chi_\Omega(x), \; x \in \mathbb{R}, \; t \in (0,T], \\ y(x,0) = y^0(x), \; y_t(x,0) = y^1(x), \qquad x \in \mathbb{R}, \end{cases} \tag{1.28}$$

*satisfies the equilibrium condition (1.24) at time $T$.*

*Proof.* The proof is similar to the one for the wave equation. It suffices to minimize the functional $\mathscr{J}$ out of the previous observability inequality, but this time within the class of solutions of the adjoint Klein–Gordon system (1.26). □

# Chapter 2
# Discontinuous Galerkin Approximations and Main Results

The second chapter of the book is twofold. First, we briefly present the approximation schemes under consideration and their main properties. In particular, we introduce the *discontinuous Galerkin* (DG) semi-discretization of the *wave* and *Klein–Gordon* equations using the so-called symmetric interior penalty DG method in its simplest version, in which *piecewise linear polynomials* are used on *uniform meshes*. We then emphasize the main difficulties encountered when analyzing the observability inequality for the DG schemes, related to the high-frequency spurious solutions propagating at small group velocities on both *physical* and *spurious* modes. In the last part of this chapter, we briefly present our main results concerning the filtering strategies we develop in this book, based on the *Fourier truncation method* and on the *bi-grid filtering technique*.

## 2.1 Discontinuous Galerkin Approximations

As we have seen in the previous chapter, observability inequalities for the $1-d$ wave equation can be easily derived out of the d'Alembert formula (1.21) and the propagation properties of solutions along characteristics. But observability and controllability issues are more subtle when the continuous wave equation is replaced by a numerical scheme. For instance, it is by now well known that the discrete analogue of the observability property (1.20) is not uniform with respect to the mesh size parameter $h$ for the classical FD and $P_1$-FEM schemes. Indeed, due to the pathological behavior of the *spurious high-frequency numerical solutions*, the discrete version of the observability constant $C(T)$ in (1.20) blows up as $h \to 0$ for any $T > 0$. The interested reader is referred to the survey articles [27, 70] for a presentation of the state of art on this topic.

The aim of this book is to extend this analysis to a class of more *sophisticated schemes*, the so-called *discontinuous Galerkin* (DG) methods.

A. Marica and E. Zuazua, *Symmetric Discontinuous Galerkin Methods for 1 – D Waves*, SpringerBriefs in Mathematics, DOI 10.1007/978-1-4614-5811-1_2,

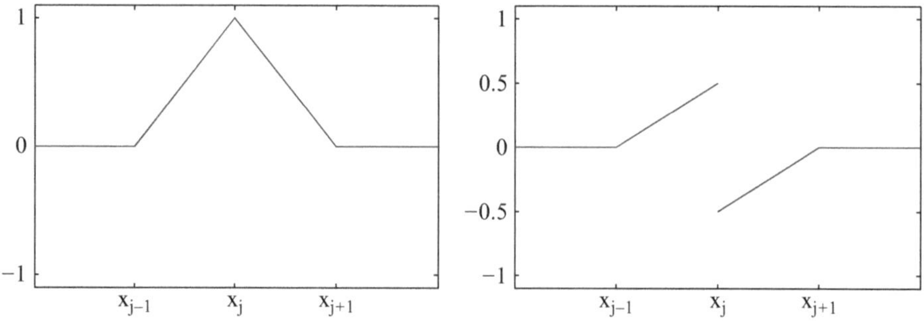


**Fig. 2.1** Basis functions for the $P_1$-DG methods: $\phi_j^{\{\cdot\}}$ (*left*) and $\phi_j^{[\cdot]}$ (*right*)

In several space dimensions, the DG methods easily handle elements of *various types* and *shapes*, *irregular nonmatching grids*, and even *varying polynomial orders*. They were proposed in the 1970s for the numerical approximation of solutions of *hyperbolic* equations and, independently, for *elliptic* and *parabolic* problems (cf. [6]). Usually, the DG approximations for *elliptic* and *parabolic* problems are called *interior penalty* (IP) methods since the *continuity* across the element interfaces is *weakly enforced* by penalizing the classical variational formulations by suitable bilinear forms, the so-called numerical fluxes. In recent years, due to their efficiency in parallel computing, intensive research has been developed on DG methods. We refer to [6, 11] for a unified analysis and comparison of the existing DG methods for *elliptic problems*.

In this book, we deal with the simplest setting of the DG space semi-discretization of the $1-d$ wave and Klein–Gordon equations with *first-order polynomials* ($P_1$) on a uniform grid $\mathcal{G}^h := \{x_j = jh, j \in \mathbb{Z}\}$ of the *whole real line* generating the *partition* (triangulation) $\mathcal{T}^h := \{I_j = (x_j, x_{j+1}), j \in \mathbb{Z}\}$ of the whole real line.

When using the $P_1$-DG method under consideration, numerical solutions can be decomposed into two essential components: *the averages* and *the jumps* of the numerical solution along the interfaces. In the $1-d$ case, the *jump* $[\cdot]$ and the *average* $\{\cdot\}$ of a function $f$ at the point $x$ are defined as $[f](x) := f(x-) - f(x+)$ and $\{f\}(x) := (f(x+) + f(x-))/2$, respectively, where $f(x\pm)$ are the right-/left-handed limits of $f$ at $x$, which may take different values when $f$ is discontinuous at $x$. Similar definitions can be given in the multidimensional case (cf. [6]).

Let $\mathcal{P}_1^h$ be the set of *polynomial functions of degree at most one* in each interval $I_j$, $j \in \mathbb{Z}$. The finite element space we consider is $\mathcal{V}^h := L^2(\mathbb{R}) \cap \mathcal{P}_1^h$. Observe that $\mathcal{V}^h$ admits the decomposition $\mathcal{V}^h = \mathcal{V}^{h,\{\cdot\}} \oplus \mathcal{V}^{h,[\cdot]}$, with $\mathcal{V}^{h,\{\cdot\}} := \text{span}\{\phi_j^{\{\cdot\}}, j \in \mathbb{Z}\}$, and $\mathcal{V}^{h,[\cdot]} := \text{span}\{\phi_j^{[\cdot]}, j \in \mathbb{Z}\}$, where (see Fig. 2.1)

$$\phi_j^{\{\cdot\}}(x) = \Big(1 - \frac{1}{h}|x - x_j|\Big)^+ \text{ and } \phi_j^{[\cdot]}(x) = \frac{1}{2}\text{sign}(x_j - x)\Big(1 - \frac{1}{h}|x - x_j|\Big)^+.$$

Here, the superscript $+$ denotes the positive part of a function.

Remark that $\phi_j^{\{\cdot\}}$ is the typical basis function used in the $P_1$-FEM, whereas $\phi_j^{[\cdot]}$, having *zero average* and *unit jump* at $x_j$, is designed to represent the *jump* at the nodal point $x_j$.

Each element $f^h \in \mathscr{V}^h$ has a unique representation as a linear combination of the form

$$f^h(x) = f^{h,\{\cdot\}}(x) + f^{h,[\cdot]}(x) := \sum_{j\in\mathbb{Z}} \{f^h\}(x_j)\phi_j^{\{\cdot\}}(x) + \sum_{j\in\mathbb{Z}} [f^h](x_j)\phi_j^{[\cdot]}(x),$$

where $f^{h,\{\cdot\}}$ and $f^{h,[\cdot]}$ are the *continuous (average)/jump* components of $f^h$, respectively. In particular, the *piecewise linear discontinuous functions* under consideration are perturbations by the *jumps* $f^{h,[\cdot]}$ added at each nodal point of the classical piecewise linear and continuous ones, represented by the *averages* $f^{h,\{\cdot\}}$.

Let us introduce some notations: the superscript $*$ is the matrix transposition. By bold lower-case characters having the superscript $h$ (e.g., $\mathbf{f}^h := (f_j)_{j\in\mathbb{Z}}$) we denote a sequence (an infinite vector) which associates to any grid point $x_j \in \mathscr{G}^h$ a unique value $f_j$. By changing the superscript $h$ into $\mathbf{h}$ (to have $\mathbf{f}^{\mathbf{h}}$), we indicate a vector associating to any grid point $x_j$ a pair of two values. In our case, any sequence $\mathbf{f}^{\mathbf{h}}$ contains two subsequences $\{\mathbf{f}^h\} := (\{f^h\}(x_j))_{j\in\mathbb{Z}}$ and $[\mathbf{f}^h] := ([f^h](x_j))_{j\in\mathbb{Z}}$, storing all the *averages/jumps* of $f^h \in \mathscr{V}^h$. All vectors under consideration are column vectors. Standard finite matrices are denoted by capital letters (e.g., $A^h$), while the bold capital letter $\mathbf{A}^h$ (or $\mathbf{A}^{\mathbf{h}}$) stands for infinite ($2\times 2$-block) matrices. Often we use *Toeplitz matrices* $\mathbf{A}^h$ with the property that all the elements along each diagonal are identical, i.e., $a_{i,j}^h = a_{i+1,j+1}^h$ for all $i, j \in \mathbb{Z}$ (cf. [34]). They arise in the numerical approximations of constant coefficients PDEs on uniform meshes. The infinite *block Toeplitz matrices* $\mathbf{A}^{\mathbf{h}}$, for which the elements on each diagonal of blocks are identical, appear, as before, in the numerical approximation of constant coefficients PDEs on uniform meshes, when using more complex methods associating to each node several degrees of freedom.

The *discrete first-order derivative* of the piecewise regular function $f^h$ having possible discontinuities only at the grid points $x_j$, $j \in \mathbb{Z}$, is denoted by $\partial_x^h f^h$ and coincides with the classical first-order derivative $\partial_x f^h$ in all the open intervals $I_j$, $j \in \mathbb{Z}$. The inner product on the space $L^2(\mathscr{T}^h)$ containing piecewise $L^2$-functions with possible discontinuities on the mesh points is denoted by $(\cdot,\cdot)_{L^2(\mathscr{T}^h)} := \sum_{j\in\mathbb{Z}}(\cdot,\cdot)_{L^2(I_j)}$. Furthermore, the inner products on the spaces $\ell^2(\mathscr{G}^h)$ and $\ell^2(\mathscr{G}^{\mathbf{h}})$ of *square summable* sequences associating *one/two* values to each grid point in $\mathscr{G}^h$ are given by

$$(\mathbf{f}^h, \mathbf{g}^h)_{\ell^2(\mathscr{G}^h)} = \sum_{j\in\mathbb{Z}} f_j \overline{g}_j, \quad (\mathbf{f}^{\mathbf{h}}, \mathbf{g}^{\mathbf{h}})_{\ell^2(\mathscr{G}^{\mathbf{h}})}$$

$$= \sum_{j\in\mathbb{Z}} (\{f^h\}(x_j)\{\overline{g}^h\}(x_j) + [f^h](x_j)[\overline{g}^h](x_j)),$$

while the corresponding norms are $||\cdot||_{L^2(\mathscr{T}^h)}$, $||\cdot||_{\ell^2(\mathscr{G}^h)}$, and $||\cdot||_{\ell^2(\mathscr{G}^{\mathbf{h}})}$.

The DG approximation we analyze in this book is the so-called *symmetric interior penalty* (SIPG) method [5, 6], whose bilinear form $\mathscr{A}_s^h : \mathscr{V}^h \times \mathscr{V}^h \to \mathbb{R}$ is defined for all *penalty parameters* $s > 1$ as follows:

$$\mathscr{A}_s^h(u^h, v^h) := (\partial_x^h u^h, \partial_x^h v^h)_{L^2(\mathscr{T}^h)} - (\{\partial_x^h \mathbf{u}^h\}, [\mathbf{v}^h])_{\ell^2(\mathscr{G}^h)} - ([\mathbf{u}^h], \{\partial_x^h \mathbf{v}^h\})_{\ell^2(\mathscr{G}^h)} + \frac{s}{h}([\mathbf{u}^h], [\mathbf{v}^h])_{\ell^2(\mathscr{G}^h)}. \tag{2.1}$$

We then consider the variational formulation of the SIPG semi-discretization of the $1-d$ wave equation

$$u_s^h(\cdot,t) \in \mathscr{V}^h, \quad (\partial_{tt} u_s^h(\cdot,t), v^h)_{L^2(\mathbb{R})} + \mathscr{A}_s^h(u_s^h(\cdot,t), v^h) = 0 \tag{2.2}$$

and of the $1-d$ Klein–Gordon equation

$$u_s^h(\cdot,t) \in \mathscr{V}^h, \quad (\partial_{tt} u_s^h(\cdot,t), v^h)_{L^2(\mathbb{R})} + \mathscr{A}_s^h(u_s^h(\cdot,t), v^h) + (u_s^h(\cdot,t), v^h)_{L^2(\mathbb{R})} = 0 \tag{2.3}$$

for all $v^h \in \mathscr{V}^h$ and $t > 0$, complemented with the initial data $u_s^h(\cdot,0) = u^{h,0} \in \mathscr{V}^h$ and $\partial_t u_s^h(\cdot,0) = u^{h,1} \in \mathscr{V}^h$.

By the Hille–Yosida theorem, there exists a unique solution of (2.2) and of (2.3), $u_s^h \in C([0,\infty), \mathscr{V}^h) \cap C^1([0,\infty), L^2(\mathbb{R}))$. Moreover, the *total energy* of solutions corresponding to the semi-discrete wave equation,

$$\mathscr{E}_s^{W,h}(\mathbf{u}^{\mathbf{h},0}, \mathbf{u}^{\mathbf{h},1}) := \frac{1}{2}\big(||\partial_t u_s^h(\cdot,t)||_{L^2} + \mathscr{A}_s^h(u_s^h(\cdot,t), u_s^h(\cdot,t))\big), \tag{2.4}$$

or to the discrete Klein–Gordon equation,

$$\mathscr{E}_s^{K,h}(\mathbf{u}^{\mathbf{h},0}, \mathbf{u}^{\mathbf{h},1}) := \frac{1}{2}\big(||\partial_t u_s^h(\cdot,t)||_{L^2}^2 + \mathscr{A}_s^h(u_s^h(\cdot,t), u_s^h(\cdot,t)) + ||u_s^h(\cdot,t)||_{L^2}^2\big), \tag{2.5}$$

is conserved in time.

The unknown $u_s^h(\cdot,t)$, being an element of $\mathscr{V}^h$ for each $t > 0$, can be decomposed as follows:

$$u_s^h(x,t) = \sum_{j\in\mathbb{Z}} \{u_s^h\}(x_j,t)\phi_j^{\{\cdot\}}(x) + \sum_{j\in\mathbb{Z}} [u_s^h](x_j,t)\phi_j^{[\cdot]}(x).$$

Using $\phi_k^{\{\cdot\}}$ and $\phi_k^{[\cdot]}$, for all $k \in \mathbb{Z}$, as test functions in (2.2) and in (2.3), we obtain that the sequence of coefficients $\mathbf{u}_s^{\mathbf{h}}(t) = (\{u_s^h\}(x_j,t), [u_s^h](x_j,t))_{j\in\mathbb{Z}}$ of the approximation $u_s^h(\cdot,t) \in \mathscr{V}^h$ is the solution of the following infinite system of second-order linear ordinary differential equations (ODEs) for the approximate wave equations for all $t > 0$:

$$\mathbf{M}^{\mathbf{h}}\partial_{tt}\mathbf{u}_s^{\mathbf{h}}(t) + \mathbf{R}_s^{\mathbf{h}}\mathbf{u}_s^{\mathbf{h}}(t) = 0, \mathbf{u}_s^{\mathbf{h}}(0) = \mathbf{u}^{\mathbf{h},0}, \ \partial_t\mathbf{u}_s^{\mathbf{h}}(0) = \mathbf{u}^{\mathbf{h},1}. \tag{2.6}$$

For the Klein–Gordon equation, we obtain the following approximating ODE system for all $t > 0$:

$$\mathbf{M^h}\partial_{tt}\mathbf{u}_s^\mathbf{h}(t) + \mathbf{R}_s^\mathbf{h}\mathbf{u}_s^\mathbf{h}(t) + \mathbf{M^h}\mathbf{u}_s^\mathbf{h}(t) = 0, \mathbf{u}_s^\mathbf{h}(0) = \mathbf{u}^{\mathbf{h},0},\ \partial_t\mathbf{u}_s^\mathbf{h}(0) = \mathbf{u}^{\mathbf{h},1}. \tag{2.7}$$

The infinite *mass* and *stiffness* matrices $\mathbf{M^h}$ and $\mathbf{R}_s^\mathbf{h}$ are *block tridiagonal*, generated by the *stencils* $M^h$ and $R_s^h$ below

$$M^h = \left(\begin{array}{cc|cc|cc} \frac{h}{6} & -\frac{h}{12} & \frac{2h}{3} & 0 & \frac{h}{6} & \frac{h}{12} \\ \frac{h}{12} & -\frac{h}{24} & 0 & \frac{h}{6} & -\frac{h}{12} & -\frac{h}{24} \end{array}\right), \quad R_s^h = \left(\begin{array}{cc|cc|cc} -\frac{1}{h} & 0 & \frac{2}{h} & 0 & -\frac{1}{h} & 0 \\ 0 & -\frac{1}{4h} & 0 & \frac{2s-1}{2h} & 0 & -\frac{1}{4h} \end{array}\right). \tag{2.8}$$

Each block in the stencils $M^h$ and $R^h$ is of dimension $2 \times 2$ since it stores the interaction between two nodes and the dynamics at each node $x_j$ is modeled by two basis functions, $\phi_j^{\{\cdot\}}$ and $\phi_j^{[\cdot]}$. Moreover, $M^h$ and $R^h$ contain three $2\times 2$-blocks since each node $x_j$ has nontrivial interactions with itself and with the two neighboring nodes, $x_{j-1}$ and $x_{j+1}$.

Observe that (2.6) or (2.7) is a coupled system of two different kinds of ODEs, each of them being generated by one of the two rows of the stencils $M^h$ and $R_s^h$.

The *total energy* $\mathscr{E}_s^{W,h}(\mathbf{u}^{\mathbf{h},0}, \mathbf{u}^{\mathbf{h},1})$ in (2.4) of the solution $\mathbf{u}_s^\mathbf{h}(t)$ of (2.6) can be written in terms of the mass and stiffness matrices $\mathbf{M^h}$ and $\mathbf{R}_s^\mathbf{h}$ as follows:

$$\mathscr{E}_s^{W,h}(\mathbf{u}^{\mathbf{h},0}, \mathbf{u}^{\mathbf{h},1}) := \frac{1}{2}(\mathbf{M^h}\partial_t\mathbf{u}_s^\mathbf{h}(t), \partial_t\mathbf{u}_s^\mathbf{h}(t))_{\ell^2(\mathscr{G}^\mathbf{h})} + \frac{1}{2}(\mathbf{R}_s^\mathbf{h}\mathbf{u}_s^\mathbf{h}(t), \mathbf{u}_s^\mathbf{h}(t))_{\ell^2(\mathscr{G}^\mathbf{h})}.$$

Also,

$$\mathscr{E}_s^{K,h}(\mathbf{u}^{\mathbf{h},0}, \mathbf{u}^{\mathbf{h},1}) := \mathscr{E}_s^{W,h}(\mathbf{u}^{\mathbf{h},0}, \mathbf{u}^{\mathbf{h},1}) + \frac{1}{2}(\mathbf{M^h}\mathbf{u}_s^\mathbf{h}(t), \mathbf{u}_s^\mathbf{h}(t))_{\ell^2(\mathscr{G}^\mathbf{h})}.$$

The aim of this book is *to analyze the propagation properties of these DG methods*. More precisely, for $\Omega := \mathbb{R} \setminus (-1, 1)$, we investigate the conditions on the observability time $T > 0$ and on the class of initial data $(\mathbf{u}^{\mathbf{h},0}, \mathbf{u}^{\mathbf{h},1})$ under consideration in (2.6) or (2.7), for which the following DG version of the *observability inequality* (1.20) holds uniformly as $h \to 0$ (here, $\varsigma \in \{W, K\}$):

$$\mathscr{E}_s^{\varsigma,h}(\mathbf{u}^{\mathbf{h},0}, \mathbf{u}^{\mathbf{h},1}) \le C_s^h(T)\int_0^T \mathscr{E}_{s,\Omega}^{\varsigma,h}(\mathbf{u}_s^\mathbf{h}(t), \partial_t\mathbf{u}_s^\mathbf{h}(t))\,dt. \tag{2.9}$$

As it occurs for classical finite difference and finite element methods, the observability inequality will fail to be uniform for the DG methods under consideration when considering all possible discrete solutions. Thus, in order to make the inequality uniform, one will need to filter the class of initial data to be considered.

The initial data $(\mathbf{u}^{\mathbf{h},0}, \mathbf{u}^{\mathbf{h},1})$ in (2.6) or (2.7) can be chosen in various different ways:

- The first one, proposed in [35], consists on taking $(\mathbf{u}^{\mathbf{h},0}, \mathbf{u}^{\mathbf{h},1})$ as the $L^2$-projection on the space $\mathscr{V}^h$ of the more regular initial data $(u^0, u^1) \in \dot{H}^{1+\sigma}(\mathbb{R}) \times \dot{H}^{\sigma}(\mathbb{R})$ in (1.18) [or $(u^0, u^1) \in H^{1+\sigma}(\mathbb{R}) \times H^{\sigma}(\mathbb{R})$ in (1.26)], with $\sigma > 0$. Namely, for each

$i = 0, 1$, the sequence $\mathbf{u}^{\mathbf{h},i}$ is the solution of the system $\mathbf{M}^{\mathbf{h}}\mathbf{u}^{\mathbf{h},i} = \mathbf{u}_{\sharp}^{\mathbf{h},i}$ where $\mathbf{u}_{\sharp}^{\mathbf{h},i} = (\{u_{\sharp}^{h,i}\}(x_j), [u_{\sharp}^{h,i}](x_j))_{j\in\mathbb{Z}}$, $\{u_{\sharp}^{h,i}\}(x_j) := (u^i, \phi_j^{\{\cdot\}})_{L^2(\mathbb{R})}$ and $[u_{\sharp}^{h,i}](x_j) := (u^i, \phi_j^{[\cdot]})_{L^2(\mathbb{R})}$. In this way, the error between the solution of the continuous model (1.18) and its DG approximation in the energy norm is of order $h^{\min(\sigma,1)}$ and the following estimates between the continuous and the discrete data hold (cf. [35]):

$$||u^1 - u^{h,1}||_{L^2}^2 \lesssim h^{2\min(\sigma,2)}||u^1||_{\dot{H}^\sigma}^2, \quad \mathscr{A}_s^h(u^0 - u^{h,0}, u^0 - u^{h,0}) \lesssim h^{2\min(\sigma,1)}||u^0||_{\dot{H}^{\sigma+1}}^2.$$

- Another possible choice for the initial data $(u^{h,0}, u^{h,1})$ in the discrete systems (2.6) and (2.7) is to take $(\{u^{h,i}\}(x_j), [u^{h,i}](x_j)) = (u^i(x_j), 0)$, which imposes *null jump conditions* at the initial time. This choice of the discrete initial data also requires $u^i$ to be *continuous* at the grid points. This condition is fulfilled for $u^0 \in \dot{H}^1(\mathbb{R}) \subset C^0(\mathbb{R})$. For the component $u^1 \in L^2(\mathbb{R})$ we can take $\{u^{h,1}\}(x_j)$ to be the average of $u^1$ on the cell $(x_{j-1/2}, x_{j+1/2})$.

For any two sequences $\mathbf{f}^{\mathbf{h},0}, \mathbf{f}^{\mathbf{h},1} \in \ell^2(\mathscr{G}^{\mathbf{h}})$, the discrete *energies concentrated in* $\Omega$, entering on the right-hand side of (2.9), are given by

$$\mathscr{E}_{s,\Omega}^{W,h}(\mathbf{f}^{\mathbf{h},0}, \mathbf{f}^{\mathbf{h},1}) := \frac{1}{2}(\mathbf{M}^{\mathbf{h}}\mathbf{f}^{\mathbf{h},1}, \mathbf{f}^{\mathbf{h},1})_{\ell^2(\mathscr{G}^{\mathbf{h}}\cap\Omega)} + \frac{1}{2}(\mathbf{R}_s^{\mathbf{h}}\mathbf{f}^{\mathbf{h},0}, \mathbf{f}^{\mathbf{h},0})_{\ell^2(\mathscr{G}^{\mathbf{h}}\cap\Omega)}$$

and

$$\mathscr{E}_{s,\Omega}^{K,h}(\mathbf{f}^{\mathbf{h},0}, \mathbf{f}^{\mathbf{h},1}) := \mathscr{E}_{s,\Omega}^{W,h}(\mathbf{f}^{\mathbf{h},0}, \mathbf{f}^{\mathbf{h},1}) + \frac{1}{2}(\mathbf{M}^{\mathbf{h}}\mathbf{f}^{\mathbf{h},0}, \mathbf{f}^{\mathbf{h},0})_{\ell^2(\mathscr{G}^{\mathbf{h}}\cap\Omega)}.$$

Here, $\ell^2(\mathscr{G}^{\mathbf{h}}\cap\Omega)$ is the space of *square summable* functions endowed with the norm $||\cdot||_{\ell^2(\mathscr{G}^{\mathbf{h}}\cap\Omega)}$ and the inner product

$$(\mathbf{f}^{\mathbf{h}}, \mathbf{g}^{\mathbf{h}})_{\ell^2(\mathscr{G}^{\mathbf{h}}\cap\Omega)} := \sum_{x_j\in\mathscr{G}^h\cap\Omega} (\{f^h\}(x_j)\{\overline{g}^h\}(x_j) + [f^h](x_j)[\overline{g}^h](x_j)).$$

## 2.2 Presentation of the Main Results

To the best of our knowledge, the present book is the first one containing a rigorous analysis of the DG methods for wave control problems, their pathologies and remedies. In what follows, we briefly describe the main results of this book.

In Chap. 3 we present a brief historical development of the DG methods under consideration and the connection between our results and the existing literature. This will allow us to compare these well-known results with the ones we shall get for the more sophisticated DG methods.

Given any sequence $\mathbf{f}^h = (f_j)_{j\in\mathbb{Z}} \in \ell^2(\mathcal{G}^h)$, we define its *semi-discrete Fourier transform* (SDFT) at scale $h$ as follows (see, e.g., [38, 67]):

$$\hat{f}^h(\xi) := h \sum_{j\in\mathbb{Z}} f_j \exp(-i\xi x_j), \text{ with } \xi \in \Pi^h := [-\pi/h, \pi/h]. \tag{2.10}$$

In what follows, by bold lower-case/capital letters with hat symbol and superscript $h$ [e.g., $\widehat{\mathbf{f}}^h(\xi)$ and $\widehat{\mathbf{F}}^h(\xi)$], we denote vectors/matrices of SDFTs, while the scalar SDFTs are represented by standard lower-case characters with hat symbol and superscript $h$ [e.g., $\hat{f}^h(\xi)$].

One of the most important properties of the SDFT at scale $h$ is its periodicity of principal period $2\pi/h$, i.e., $\hat{f}^h(\xi) = \hat{f}^h(\xi+2k\pi/h)$ for all $\xi \in \Pi^h$ and $k \in \mathbb{Z}$. This explains the choice of the domain of definition for the SDFT to be the interval $\Pi^h$ of length $2\pi/h$. Although any interval of the form $[\xi^0, \xi^0 + 2\pi/h]$, with $\xi^0 \in \mathbb{R}$, could be chosen as domain of definition for the SDFT, $\Pi^h$ is the only possible interval of the above form which is symmetric with respect to $\xi = 0$, a property that the domain $\mathbb{R}$ of the continuous Fourier transform also has. In fact, as $h \to 0$, the interval $\Pi^h$ covers the whole real line.

The second important property of the SDFT is the *Parseval identity*:

$$||\mathbf{f}^h||^2_{\ell^2(\mathcal{G}^h)} = \frac{1}{2\pi}||\hat{f}^h||^2_{L^2(\Pi^h)}, \quad \forall \mathbf{f}^h \in \ell^2(\mathcal{G}^h). \tag{2.11}$$

The third property of the SDFT useful for our analysis is the fact that the SDFT can be inverted. Indeed, for any $\mathbf{f}^h \in \ell^2(\mathcal{G}^h)$ and for any $j \in \mathbb{Z}$, the following identity holds:

$$f_j = \frac{1}{2\pi}\int_{\Pi^h} \hat{f}^h(\xi)\exp(i\xi x_j)\, d\xi. \tag{2.12}$$

Chapter 4 is devoted to a careful *Fourier analysis* of the systems (2.6) and (2.7). Their solutions $\mathbf{u}^{\mathbf{h}}_s(t)$ can also be written as a pair $(\{\mathbf{u}^h_s(\cdot,t)\}, [\mathbf{u}^h_s(\cdot,t)])$ composed by the two sequences of different nature containing the *averages/jumps* of the numerical solutions. Let $\widehat{\mathbf{u}}^h_s(\xi,t) := (\hat{u}^{h,\{\cdot\}}_s(\xi,t), \hat{u}^{h,[\cdot]}_s(\xi,t))$ be the column vector of the SDFTs of the sequence of *averages* $\{\mathbf{u}^h_s(\cdot,t)\}$ and of *jumps* $[\mathbf{u}^h_s(\cdot,t)]$. When dealing with system (2.6), the vector function $\widehat{\mathbf{u}}^h_s(\xi,t)$ satisfies the following system of second-order ODEs in time, depending on the wave number parameter $\xi \in \Pi^h$:

$$\begin{cases} \widehat{\mathbf{M}}^h(\xi)\partial_{tt}\widehat{\mathbf{u}}^h_s(\xi,t) + \widehat{\mathbf{R}}^h_s(\xi)\widehat{\mathbf{u}}^h_s(\xi,t) = 0, \\ \widehat{\mathbf{u}}^h_s(\xi,0) = \widehat{\mathbf{u}}^{h,0}(\xi),\ \partial_t\widehat{\mathbf{u}}^h_s(\xi,0) = \widehat{\mathbf{u}}^{h,1}(\xi), \end{cases} \tag{2.13}$$

where $\xi \in \Pi^h$, $t > 0$, and $\widehat{\mathbf{M}}^h(\xi)$ and $\widehat{\mathbf{R}}^h_s(\xi)$ are the matrix Fourier symbols of the mass and stiffness matrices $\mathbf{M}^{\mathbf{h}}$ and $\mathbf{R}^{\mathbf{h}}_s$ given by

$$\widehat{\mathbf{M}}^h(\xi) = \begin{pmatrix} \frac{2+\cos(\xi h)}{3} & \frac{i\sin(\xi h)}{6} \\ -\frac{i\sin(\xi h)}{6} & \frac{2-\cos(\xi h)}{12} \end{pmatrix},\ \widehat{\mathbf{R}}^h_s(\xi) = \begin{pmatrix} \frac{4}{h^2}\sin^2\left(\frac{\xi h}{2}\right) & 0 \\ 0 & \frac{s-\cos^2\left(\frac{\xi h}{2}\right)}{h^2} \end{pmatrix}. \tag{2.14}$$

Using the *Parseval identity* (2.11), we find the following *Fourier representation* of the *total energy* (2.4) (in which $(\cdot,\cdot)_{\mathbb{C}^n}$ is the inner product in $\mathbb{C}^n$, with $n \in \mathbb{N}$):

$$\mathscr{E}_s^{W,h}(\mathbf{u}^{\mathbf{h},0}, \mathbf{u}^{\mathbf{h},1})$$
$$= \frac{1}{4\pi} \int_{\Pi^h} \left[ (\widehat{\mathbf{R}}_s^h(\xi)\widehat{\mathbf{u}}^{h,0}(\xi), \widehat{\mathbf{u}}^{h,0}(\xi))_{\mathbb{C}^2} + (\widehat{\mathbf{M}}^h(\xi)\widehat{\mathbf{u}}^{h,1}(\xi), \widehat{\mathbf{u}}^{h,1}(\xi))_{\mathbb{C}^2} \right] d\xi. \quad (2.15)$$

Remark that, unlike (2.6), which is an infinite system, (2.13) can be explicitly solved in terms of the *eigenvalues* $\hat{\Lambda}_s^h(\xi)$ of the matrix

$$\widehat{\mathbf{S}}_s^h(\xi) := (\widehat{\mathbf{M}}^h(\xi))^{-1}\widehat{\mathbf{R}}_s^h(\xi) \quad (2.16)$$

and of its *eigenvectors* $\widehat{\mathbf{v}}_s^h(\xi)$. The two families of eigensolutions $(\hat{\Lambda}_s^h(\xi), \widehat{\mathbf{v}}_s^h(\xi))$ of the matrix $\widehat{\mathbf{S}}_s^h(\xi)$ are denoted by $(\hat{\Lambda}_{s,\alpha}^h(\xi), \widehat{\mathbf{v}}_{s,\alpha}^h(\xi))$, $\alpha \in \{\text{ph}, \text{sp}\}$, where the subscripts ph/sp stand for *physical/spurious*. Set

$$\hat{\lambda}_{s,\alpha}^h(\xi) := \text{sign}(\xi)\sqrt{\hat{\Lambda}_{s,\alpha}^h(\xi)}, \quad (2.17)$$

with $\alpha \in \{\text{ph}, \text{sp}\}$, to be the so-called dispersion relations of the SIPG method under consideration.

As we can see in Fig. 2.2 and rigorously in Chap. 4, the *physical dispersion relation* $\hat{\lambda}_{s,\text{ph}}^h(\xi)$ lies between the two diagrams corresponding to the FD and $P_1$-FEM semi-discretizations, $\hat{\lambda}_{1,\text{ph}}^h(\xi)$ and $\hat{\lambda}_{\infty,\text{ph}}^h(\xi)$. The *spurious dispersion relation*, $\hat{\lambda}_{s,\text{sp}}^h(\xi)$, tends to infinity for large values of $s$. One can prove that, for all $s \in (1,\infty) \setminus \{3\}$, the physical dispersion relation is strictly increasing for all the wave numbers $\xi \in (0, \pi/h)$, with only one critical point located at $\xi = \pi/h$, at which its first-order derivative $\partial_\xi \hat{\lambda}_{s,\text{ph}}^h(\xi)$, the so-called physical group velocity, vanishes. Concerning the monotonicity of the spurious dispersion relation, several ranges of $s$ can be identified according to the *stabilization parameter* $s$. Excepting for the values $s \in (5/3, 5/2)$, for which there are three critical points on $\hat{\lambda}_{s,\text{sp}}^h(\xi)$, and $s = 3$, for which there exists only the critical point $\xi = 0$, for all the other values of $s$, the *group velocity* $\partial_\xi \hat{\lambda}_{s,\text{sp}}^h(\xi)$ vanishes at two wave numbers: $\xi = 0$ and $\xi = \pi/h$.

A similar Fourier analysis to the one in Chap. 4 can be also applied to the discrete Klein–Gordon equation (2.7). By taking SDFT in the approximate Klein–Gordon equation (2.7), we obtain

$$\begin{cases} \widehat{\mathbf{M}}^h(\xi)\partial_{tt}\widehat{\mathbf{u}}_s^h(\xi,t) + \widehat{\mathbf{R}}_s^h(\xi)\widehat{\mathbf{u}}_s^h(\xi,t) + \widehat{\mathbf{M}}^h(\xi)\widehat{\mathbf{u}}_s^h(\xi,t) = 0, \\ \widehat{\mathbf{u}}_s^h(\xi,0) = \widehat{\mathbf{u}}^{h,0}(\xi), \ \partial_t\widehat{\mathbf{u}}_s^h(\xi,0) = \widehat{\mathbf{u}}^{h,1}(\xi), \end{cases} \quad (2.18)$$

so that the two eigenvalues of matrix $(\widehat{\mathbf{M}}^h(\xi))^{-1}\widehat{\mathbf{S}}_s^h(\xi) + I_2$ corresponding to system (2.18) are one unit higher than the ones of matrix $(\widehat{\mathbf{M}}^h(\xi))^{-1}\widehat{\mathbf{S}}_s^h(\xi)$ in system (2.13), while the eigenvectors are the same.

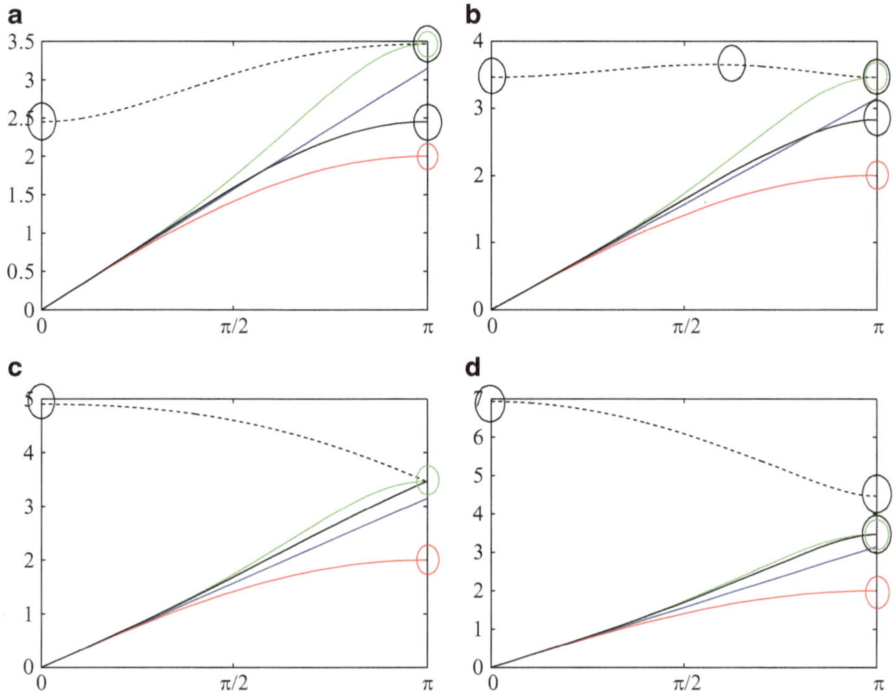


**Fig. 2.2** The physical/spurious dispersion relation for the SIPG method, $\hat{\lambda}^1_{s,\alpha}(\xi)$, $\alpha = \mathrm{ph}/\mathrm{sp}$ (*black/dotted black*), compared to the ones of the continuous wave equation $\xi$ (*blue*) and of its FD and $P_1$-FEM semi-discretizations, $\hat{\lambda}^1_{1,\mathrm{ph}}(\xi)$ (*red*) and $\hat{\lambda}^1_{\infty,\mathrm{ph}}(\xi)$ (*green*). The marked points are wave numbers where the corresponding group velocities vanish (**a**) $s = 1.5$ (**b**) $s = 2$ (**c**) $s = 3$ (**d**) $s = 5$

Similarly to (2.15), the *Fourier representation* of the *total energy* (2.5) is

$$\mathscr{E}_s^{K,h}(\mathbf{u}^{\mathbf{h},0}, \mathbf{u}^{\mathbf{h},1}) = \mathscr{E}_s^{W,h}(\mathbf{u}^{\mathbf{h},0}, \mathbf{u}^{\mathbf{h},1}) + \frac{1}{4\pi}\int_{\Pi^h} (\widehat{\mathbf{M}}^h(\xi)\widehat{\mathbf{u}}^{h,0}(\xi), \widehat{\mathbf{u}}^{h,0}(\xi))_{\mathbb{C}^2}\, d\xi. \tag{2.19}$$

The existence of wave numbers $\xi$, where one of the two *group velocities* (the *physical* or the *spurious* one) vanishes, necessarily implies that the observability constant $C_s^h(T)$ in (2.9) blows up at least polynomially at any order as the mesh size parameter $h$ tends to zero. The behavior of the constant $C_s^h(T)$ will be analyzed in Chap. 5 by adapting well-known constructions of *high-frequency Gaussian wave packets* previously implemented for the classical approximations (FD or $P_1$-FEM) in [28] or [49].

Consequently, the effect of these pathological wave numbers and components of the dispersion diagram need to be attenuated or filtered out. This needs to be done, in particular, with the high-frequency components at $\xi = \pi/h$ on the physical dispersion diagram or the ones at $\xi = 0$ and at $\xi = \pi/h$ on the spurious one. Chapter 6 is devoted to designing and analyzing appropriate *filtering techniques* for the SIPG method aimed to face the pathologies introduced by the singularities of the two dispersion diagrams, i.e., to reestablish the discrete observability inequality (2.9) with a constant $C_s^h(T)$ independent of $h$. All these filtering mechanisms consist in fact in imposing restrictions on the initial data $(\mathbf{u^{h,0}}, \mathbf{u^{h,1}})$ in (2.6) so that the observability result (2.9) is uniform with respect to $h$. Due to the complex structure of the Fourier decomposition of the systems under consideration, each one of these filtering algorithms combines two techniques previously proved to be efficient for the FD or the $P_1$-FEM approximations. Let us briefly present the four filtering strategies we analyze.

**Algorithm A**. The algorithm in Sect. 6.1 is based on the use of numerical initial data in (2.6) or (2.7) involving only the physical mode and chosen to be of the form $\widehat{\mathbf{u}}^{h,i}(\xi) := \widehat{\mathbf{v}}^h_{s,\text{ph}}(\xi)\hat{u}^{h,i}(\xi)$ for all $i = 0, 1$, where $\widehat{\mathbf{v}}^h_{s,\text{ph}}(\xi)$ is the *physical eigenvector* of the SIPG method and $\hat{u}^{h,i}(\xi)$ is a scalar function. This choice yields solutions of (2.13) involving only the *physical mode*. Then we apply the *Fourier truncation method*, so that the support of the Fourier transform $(\hat{u}^{h,0}(\xi), \hat{u}^{h,1}(\xi))$ of the initial data is contained in $\Pi^h_\delta := [-\pi\delta/h, \pi\delta/h]$ for some $\delta \in (0,1)$ and it does not contain the critical point $\pi/h$ of the physical dispersion relation.

**Algorithm B**. The strategy in Sect. 6.2 is also based on data involving only the physical mode as in Algorithm A, but with $(\hat{u}^{h,0}(\xi), \hat{u}^{h,1}(\xi))$ obtained as SDFTs of two sequences $\mathbf{u}^{h,0}$ and $\mathbf{u}^{h,1}$ produced by the classical *bi-grid algorithm*, i.e., only its odd components $(u^i_{2j+1})_{j\in\mathbb{Z}}$ are given, while the even ones are defined as *linear interpolation* of the two neighboring odd ones. We refer to [37, 59] for the basic properties of data in the bi-grid class.

This algorithm is of purely theoretical interest, since the practical utility of the bi-grid algorithm consists precisely in the fact that it does not need to use the Fourier transform and its inverse to compute the numerical approximation, but rather that it can be fully implemented on the physical grid. The goal of our analysis in Sect. 6.2 is to highlight that, at least from a theoretical point of view, once a dispersion relation having only a critical point at $\xi = \pi/h$ is generated, independently of the complexity of its formula, the bi-grid algorithm is still efficient to attenuate the high-frequency pathologies of the underlying solutions.

**Algorithm C**. The filtering algorithm in Sect. 6.3 is based on the use of initial data with null jump components, i.e., their *jump* components $[\mathbf{u}^{h,i}]$ vanish for both $i = 0$ and $i = 1$. The *average* components $\{\mathbf{u}^{h,i}\}$ are then given by the *Fourier truncation method* with parameter $\delta \in (0,1)$ as in the Algorithm A, to avoid the critical point on the physical diagram at $\xi = \pi/h$.

**Algorithm D**. The strategy in Sect. 6.4 is also based on initial data with null jump components as in Algorithm C, but now the averages $\{\mathbf{u}^{h,i}\}$ are defined by a *bi-grid algorithm* of mesh ratio $1/2$, as in Algorithm B.

Chapter 7 is devoted to briefly analyze some other closely related methods such as the *classical quadratic finite element* method ($P_2$-FEM), the so-called $P_1$-*local discontinuous Galerkin* ($P_1$-LDG) method and a version of the SIPG method in which both the jumps of the numerical solution and of its normal derivative along the interfaces are penalized (SIPG-n). As we shall see, the analysis of all these methods can be developed using the tools presented in this book.

# Chapter 3
# Bibliographical Notes

The aim of this chapter is to present a brief overview on the evolution of the DG methods under consideration and especially of their use in problems related to wave propagation and control. However, our presentation is far from being exhaustive and the interested reader is oriented to more extensive review papers (e.g., [17] and [19]). We also describe the influence of this historical evolution on our work.

The SIPG method we use was introduced by Arnold in 1982 (cf. [5]), while the LDG method was introduced by Cockburn–Shu in 1998 (cf. [20]). In 2002, Arnold–Brezzi–Cockburn–Marini put all the existing DG methods for elliptic problems in a unified framework, in which general numerical fluxes are considered and the different DG methods are obtained by particularizing them (cf. [6]).

In [2, 7, 36] and [61], Ainsworth, Atkins–Hu, Hu–Hussaini–Rasetarinera, and Sherwin performed a *plane wave analysis* of the DG semi-discretizations of the $1-d$ transport equation on uniform meshes obtained by using the *Lax–Friedrich numerical fluxes*. It was proved that for any polynomial degree of the numerical approximation, two numerical waves (a *physical/spurious* one propagating in the *correct/nonphysical* direction) emerge for each continuous-time periodic monochromatic wave, except for the particular case of the *exact characteristics splitting flux formula* for which there exists only one numerical wave propagating in the right direction. Thus, each given real frequency $\omega$ generates at most two *complex* discrete wave numbers: a *physical* one, $\xi^h_{ph}$, and a *spurious* one, $\xi^h_{sp}$, of real parts of opposite signs. In our analysis, for any fixed *real* wave number $\xi \in \Pi^h$, we obtain two *real* frequencies $\hat{\lambda}^h_{s,ph}(\xi)$ and $\hat{\lambda}^h_{s,sp}(\xi)$. In [2], the quantity of interest is the *phase velocity*, giving information about the propagation of monochromatic waves, while we focus on the *group velocity* (giving information about the propagation of *wave packets*). The two phase velocities $\hat{\lambda}^h_{s,\alpha}(\xi)/\xi$, with $\alpha \in \{ph, sp\}$, have the same sign, while the two group velocities $\partial_\xi \hat{\lambda}^h_{s,\alpha}(\xi)$ typically have opposite signs.

In 2006, Antonietti–Buffa–Perugia (cf. [4]) have considered the problem of *computing the eigensolutions* of the Laplace operator approximated by means of DG methods in *several space dimensions* and on general *nonuniform triangulations*.

A. Marica and E. Zuazua, *Symmetric Discontinuous Galerkin Methods for 1 – D Waves*, SpringerBriefs in Mathematics, DOI 10.1007/978-1-4614-5811-1_3,

They show that some DG methods provide *spectrally correct* approximations (the SIPG and LDG methods under consideration belong to this class). Roughly speaking, this means that the low-frequency part of the physical mode $\hat{\lambda}^h_{s,ph}(\xi)$ approximates well the continuous symbol $\xi^2$, for all $\xi << h^{-\alpha}$ and some $\alpha \in (0,2)$. However, no description concerning the behavior of the eigenvalues of the discrete Laplacian in the high-frequency regime is provided, while our analysis requires fine properties of the whole spectrum that we are able to obtain in a *simplified setting* (i.e., *one* space dimension and *uniform* meshes on the whole real line, so that *no boundary conditions* are imposed).

In [3], Ainsworth–Monk–Muniz have shown that the bilinear form $\mathscr{A}^h_s$ in (2.1) associated to the SIPG method on $1-d$ uniform meshes is coercive if the penalty parameter $s$ satisfies the lower bound $s > p(p+1)/2$. For $p = 1$, this yields precisely the values $s > 1$ we consider in this book. Moreover, the *dispersion error* $|\hat{\lambda}^h_{s,ph}(\xi) - \xi|$ is proved to be of order $(\xi h)^{2p+1}/h$ for (a) any odd $p$ and any $s > p(p+1)/2$ and for (b) any even $p$ and $s \in (p(p+1)/2, \infty)\backslash\{(p+1)(p+2)/2\}$ and of order $(\xi h)^{2p+3}/h$ for c) any even $p$ and $s = (p+1)(p+2)/2$. Thus, excepting the case c), the dispersion error for the SIPG method is the same as the one for the classical FEM approximations. In particular, this means that for any value of $s$, the $P_1$-SIPG approximation of the control problem (1.28) does not provide better results than the FD or the $P_1$-FEM in what concerns the approximation order of the numerical controls (cf. [27]).

In 2006, Grote–Schneebeli–Schötzau have shown that the SIPG semi-discrete wave equation we use is *convergent* in the *classical sense* of numerical analysis (cf. [35]). More precisely, optimal *a priori error bounds* of order $h^{\min\{\sigma,p\}}$ are obtained in the energy norm, where $\sigma$ is the regularity of the initial data in the continuous problem (1.18) (i.e., $(u^0,u^1) \in \dot{H}^{1+\sigma} \times \dot{H}^{\sigma}(\mathbb{R})$).

Note however that, as indicated in the introduction, our analysis concerns not only the numerical solutions with fixed initial data but also, especially, the behavior of numerical solutions for oscillatory data of wavelength of order $h$.

The recent work [55] of Mariegaard (2009) devoted to both the conforming $P_1$-FEM and a DG approximation for the boundary control of the $1-d$ wave equation is also worth mentioning. With respect to our approach, which consists in replacing the continuous Laplacian by an *interior penalty* DG method, in [55], the continuous d'Alembert operator (1.18) is decomposed into two transport operators, each of them being approximated by means of a DG method for conservation laws with *Lax–Friedrich* numerical fluxes. A numerical study of the Gramian operator $\int_0^T \exp(it\sqrt{\Delta^h})B^h(B^h)^* \exp(-it\sqrt{\Delta^h})\,dt$ was performed, $\Delta^h$ being a discrete version of the Laplace operator and $B^h$ the discretization of the boundary observation operator. In the continuous case, for $T = 2$, the Gramian operator is the unit matrix. For the $P_1$-FEM approximation of the wave equation, the discrete Gramian has also a diagonal structure with the maximal elements concentrated along the main diagonal, whose values are going from one (for the low-frequency components) to zero (for the high-frequency ones), which causes its pathological behavior. For the $P_1$-DG approximation, the discrete Gramian has a two-diagonal structure given by the main diagonal (corresponding to the physical mode) and a parallel to the anti-diagonal (corresponding to the spurious mode). The efficiency of the *Fourier truncation method* on the structure of the Gramian matrix was also investigated.

In 2010, Agut–Diaz (cf. [1]) have studied the dependence on $s$ and $p$ of the *Courant–Friedrich–Lewy* (CFL) condition for the fully discrete versions of the SIPG approximations of the wave equation using the *leapfrog* explicit scheme as time discretization. As a general rule, the *Courant number* $\mu := \delta t/h$ has the upper bound $\mu \leq 2/\max|\hat{\lambda}_s^1(\xi)|$, where $\delta t$ is the time step, $\xi \in \Pi^1$, and $\hat{\lambda}_s^h(\xi)$ is any dispersion relation of the SIPG method. The results in [1] for the simplest case when $p = 1$ can be easily explained in view of the analysis of the monotonicity with respect to $\xi$ of the two dispersion relations $\hat{\lambda}^h_{s,ph}(\xi)$ and $\hat{\lambda}^h_{s,sp}(\xi)$ we presented before. Thus, for $s \in (1, 5/3)$, the upper bound of $\mu$, $1/\sqrt{3}$, is constant with respect to $s$; for $s \in (5/3, 5/2)$, due to the change of monotonicity of the spurious diagram, the CFL condition has a smooth transition region from $1/\sqrt{3}$ to $\sqrt{2}/3$; for $s > 5/2$, the CFL condition decreases with respect to $s$ like $1/\sqrt{3(s-1)}$. This makes the numerical computations unfeasible for large values of $s$ when the spurious diagram is not filtered out. One can also use *conservative implicit time-discretization schemes* to avoid any restriction on $\mu$ (cf. [25]).

# Chapter 4
# Fourier Analysis of the Discontinuous Galerkin Methods

The purpose of this chapter is to perform a complete Fourier analysis of the discontinuous Galerkin method under consideration. We obtain explicit formulas for the two classes of eigenvalues and eigenvectors of the Fourier symbol of the discrete Laplacian, the so-called *physical* and *spurious* modes. In the last part of this chapter, we analyze fine properties of both Fourier modes as, for example, the behavior of the corresponding *group velocities* which are first-order derivatives of the square roots of the eigenvalues of the discrete Laplacian. The *group velocity* is an important notion in wave propagation. In particular, as we will explain within the next chapter, the existence of critical wave numbers where one of the two group velocities vanishes yields wave packets with null velocity of propagation.

More precisely, our objective is to analyze fine properties of the eigensolutions $(\hat{\Lambda}^h_{s,\alpha}(\xi), \hat{\mathbf{v}}^h_{s,\alpha}(\xi))$, with $\alpha \in \{ph, sp\}$, of the matrix $\hat{\mathbf{S}}^h_s(\xi)$ in (2.16), given explicitly as follows:

$$\hat{\mathbf{S}}^h_s(\xi) := \begin{pmatrix} (2-\cos(\xi h))\frac{4}{h^2}\sin^2\left(\frac{\xi h}{2}\right) & -2i\sin(\xi h)\frac{s-\cos^2\left(\frac{\xi h}{2}\right)}{h^2} \\ 2i\sin(\xi h)\frac{4}{h^2}\sin^2\left(\frac{\xi h}{2}\right) & 4(2+\cos(\xi h))\frac{s-\cos^2\left(\frac{\xi h}{2}\right)}{h^2} \end{pmatrix}.$$

## 4.1 Analysis of the Spectrum of the SIPG Method

Since $\hat{\mathbf{S}}^h_s(\xi)$ is a $2\times 2$-matrix, its eigenvalues $\hat{\Lambda} = \hat{\Lambda}^h_s(\xi)$ satisfy the second-order algebraic equation

$$\hat{\Lambda}^2 - \frac{2\hat{\Lambda}}{h^2}[12+2(s-3)(2+\cos(\xi h))] + \frac{48}{h^4}\sin^2\left(\frac{\xi h}{2}\right)\left[s-\cos^2\left(\frac{\xi h}{2}\right)\right] = 0. \quad (4.1)$$

A. Marica and E. Zuazua, *Symmetric Discontinuous Galerkin Methods for 1 – D Waves*, SpringerBriefs in Mathematics, DOI 10.1007/978-1-4614-5811-1_4,

Set the *discriminant* of (4.1) to be

$$\hat{\Delta}_s^h(\xi) := (12 + 2(s-3)(2+\cos(\xi h)))^2 - 48\sin^2\left(\frac{\xi h}{2}\right)\left(s - \cos^2\left(\frac{\xi h}{2}\right)\right). \tag{4.2}$$

The two eigenvalues $\hat{\Lambda} = \hat{\Lambda}_s^h(\xi)$ of $\hat{\mathbf{S}}_s^h(\xi)$ are explicitly given by

$$\hat{\Lambda}_{s,\alpha}^h(\xi) := \frac{1}{h^2}\left(12 + 2(s-3)(2+\cos(\xi h)) + \text{sign}(\alpha)\sqrt{\hat{\Delta}_s^h(\xi)}\right),$$

where $\alpha \in \{ph, sp\}$, $\text{sign}(ph) = -1$ and $\text{sign}(sp) = 1$. Following the terminology in [10] and [12], we refer to $\hat{\Lambda}_{s,ph}^h(\xi)$ and $\hat{\Lambda}_{s,sp}^h(\xi)$ as the *physical (acoustic)/spurious (optic) Fourier symbols*, respectively.

In what follows, we will use several expressions involving negative powers of $\hat{\Delta}_s^h(\xi)$. For this reason, we need the following result describing the points $(\xi, s) \in \Pi^h \times (1, \infty)$ where $\hat{\Delta}_s^h(\xi)$ vanishes.

**Lemma 4.1.** *The discriminant $\hat{\Delta}_s^h(\xi)$ has the following property:*

$$\hat{\Delta}_s^h(\xi) = 0 \text{ iff } \left(\xi = \pm\pi/h \text{ and } s = 3\right) \text{ or } \left(\xi = 0 \text{ and } s = 1\right).$$

*Otherwise, $\hat{\Delta}_s^h(\xi) > 0$.*

However, in order to ensure the stability of the scheme, we will focus on the range of parameters $s > 1$.

*Remark 4.1.* Lemma 4.1 has two important consequences on our analysis:

- The physical and spurious branches of the spectrum are symmetric with respect to the curve $12 + 2(s-3)(2+\cos(\xi))$, being situated at the distance $\sqrt{\hat{\Delta}_s^1(\xi)}$ from it. The fact that for $s > 1$ there exists only one value of $(\xi, s) = (\pi, 3)$ for which $\hat{\Delta}_s^1(\xi)$ vanishes implies that, for $s \neq 3$, the two Fourier symbols are well separated, while for $s = 3$ they have a unique bifurcation point at $\xi = \pi$ (both eigenvalues are even functions, so that here we refer only to their behavior for $\xi \in [0, \pi]$).
- From (4.13), we observe that two of the factors involved in both group velocities $\partial_\xi \hat{\lambda}_{s,\alpha}^h(\xi)$, with $\alpha \in \{ph, sp\}$, are $\cos(\xi h/2)$ and $1/\sqrt{\hat{\Delta}_s^h(\xi)}$. Thus, for $s \neq 3$, both group velocities vanish at $\xi = \pi/h$ since $\hat{\Delta}_s^h(\xi) > 0$ for all $\xi \in [0, \pi/h]$. Nevertheless, for $s = 3$, it can be proved that $\hat{\Delta}_s^h(\xi)$ vanishes like $\cos^2(\xi h/2)$ as $\xi \to \pi/h$, so that none of the group velocities vanishes at $\xi = \pi/h$.

*Proof of Lemma 4.1.* Some easy computations show that $\hat{\Delta}_s^h(\xi)$ can be written as a sum of two nonnegative quantities as follows:

$$\hat{\Delta}_s^h(\xi) = \left(2(s-3)(2+\cos(\xi h)) + \frac{36\cos^2\left(\frac{\xi h}{2}\right)}{2+\cos(\xi h)}\right)^2 + \frac{192\cos^2\left(\frac{\xi h}{2}\right)\sin^6\left(\frac{\xi h}{2}\right)}{(2+\cos(\xi h))^2}. \tag{4.3}$$

The last term in (4.3) vanishes only in the following two cases: (a) $\cos(\xi h/2) = 0$, for which the first term in (4.3) vanishes if $s = 3$; (b) $\sin(\xi h/2) = 0$, for which the first term in (4.3) vanishes if $s = 1$. This concludes the proof of Lemma 4.1. □

For any fixed $\xi \in \Pi^h$, we have

$$\hat{\Lambda}^h_{s,ph}(\xi) \to \hat{\Lambda}^h_{\infty,ph}(\xi) \text{ as } s \to \infty$$

and

$$\hat{\Lambda}^h_{s,ph}(\xi) \to \hat{\Lambda}^h_{1,ph}(\xi) \text{ as } s \to 1,$$

where $\hat{\Lambda}^h_{\infty,ph}(\xi)$ and $\hat{\Lambda}^h_{1,ph}(\xi)$ are the Fourier symbols of the $P_1$-FEM and FD approximations introduced in (1.17). Moreover, the following result concerning the comparison between the *physical symbol* of the SIPG approximation, $\hat{\Lambda}^h_{s,ph}(\xi)$, and the ones for the FD and $P_1$-FEM approximations, $\hat{\Lambda}^h_{1,ph}(\xi)$ and $\hat{\Lambda}^h_{\infty,ph}$, holds (see also Fig. 2.2):

**Lemma 4.2.** *For each $s' > s \geqslant 1$ and $\xi \in \Pi^h$, the following inequalities hold:*

$$\hat{\Lambda}^h_{1,ph}(\xi) \leqslant \hat{\Lambda}^h_{s,ph}(\xi) \leqslant \hat{\Lambda}^h_{s',ph}(\xi) \leq \hat{\Lambda}^h_{\infty,ph}(\xi). \tag{4.4}$$

*Proof.* First, we prove the last inequality in (4.4). It is easy to check that

$$\hat{\Delta}^h_{s'}(\xi) \geq \left|4(2+\cos(\xi h))\left(s' - \cos^2\left(\frac{\xi h}{2}\right)\right) - \left(12 + 2(s'-3)(2+\cos(\xi h))\right)\right|^2,$$

which implies that

$$\frac{\hat{\Lambda}^h_{s',ph}(\xi)}{3\hat{\Lambda}^h_{1,ph}(\xi)} = \frac{4\left(s' - \cos^2\left(\frac{\xi h}{2}\right)\right)}{12 + 2(s'-3)(2+\cos(\xi h)) + \sqrt{\hat{\Delta}^h_{s'}(\xi)}} \leqslant \frac{1}{2+\cos(\xi h)} = \frac{\hat{\Lambda}^h_{\infty,ph}(\xi)}{3\hat{\Lambda}^h_{1,ph}(\xi)}. \tag{4.5}$$

From (4.5) we obtain the last inequality in (4.4). By using it, we get

$$\partial_s \hat{\Lambda}^h_{s,ph}(\xi) = \frac{2(2+\cos(\xi h))}{\sqrt{\Delta^s_h(\xi)}}\left[\hat{\Lambda}^h_{\infty,ph}(\xi) - \hat{\Lambda}^h_{s,ph}(\xi)\right] \geqslant 0, \tag{4.6}$$

so that $\hat{\Lambda}^h_{s,ph}(\xi)$ is increasing with respect to $s$ for all $\xi \in [0, \pi/h]$. Its maximum value, $\hat{\Lambda}^h_{\infty,ph}(\xi)$, corresponding to the $P_1$-FEM, is obtained as $s \to \infty$, and the minimal one, $\hat{\Lambda}^h_{1,ph}(\xi)$, corresponding to the FD scheme, is attained for $s = 1$. This concludes the proof of (4.4). □

At the same time, for fixed $h > 0$, the spurious symbol $\hat{\Lambda}^h_{s,sp}(\xi)$ tends to infinity as $s \to \infty$.

The two eigenvectors of the matrix $\hat{\mathbf{S}}^h_s(\xi)$ are:

$$\hat{\mathbf{v}}^h_{s,ph}(\xi) = \frac{1}{\sqrt{1 + |\hat{v}^h_{s,ph}(\xi)|^2}} \begin{pmatrix} 1 \\ \hat{v}^h_{s,ph}(\xi) \end{pmatrix} \tag{4.7}$$

and

$$\hat{\mathbf{v}}^h_{s,sp}(\xi) = \frac{1}{\sqrt{1 + |\hat{v}^h_{s,sp}(\xi)|^2}} \begin{pmatrix} \hat{v}^h_{s,sp}(\xi) \\ 1 \end{pmatrix},$$

where

$$\hat{v}^h_{s,ph}(\xi) := \frac{4 \sin^2\left(\frac{\xi h}{2}\right)}{s - \cos^2\left(\frac{\xi h}{2}\right)} \hat{v}^h_{s,sp}(\xi) \tag{4.8}$$

and

$$\hat{v}^h_{s,sp}(\xi) := \frac{1}{i} \frac{1}{2\sin(\xi h)} \left( 2 - \cos(\xi h) - \frac{12\left(s - \cos^2\left(\frac{\xi h}{2}\right)\right)}{h^2 \hat{\Lambda}^h_{s,sp}(\xi)} \right).$$

The following two propositions describe the behavior of the two eigenvectors $\hat{\mathbf{v}}^h_{s,ph}(\xi)$ and $\hat{\mathbf{v}}^h_{s,sp}(\xi)$.

**Proposition 4.1.** *The physical eigenvector $\hat{\mathbf{v}}^h_{s,ph}(\xi)$ has the following properties (see Fig. 4.1):*

*(a1)* $\lim_{\xi \to 0} \hat{v}^h_{s,ph}(\xi) = 0$ *and* $\lim_{\xi \to 0} \hat{\mathbf{v}}^h_{s,ph}(\xi) = (1 \quad 0)^*$ *for all* $s > 1$.

*(a2)* $\lim_{\xi \to \pm\pi/h} \hat{v}^h_{s,ph}(\xi) = 0$ *and* $\lim_{\xi \to \pm\pi/h} \hat{\mathbf{v}}^h_{s,ph}(\xi) = (1 \quad 0)^*$ *for all* $s > 3$.

*(a3)* $\lim_{\xi \to \pm\pi/h} \hat{v}^h_{3,ph}(\xi) = \pm\frac{2}{i\sqrt{3}}$ *and* $\lim_{\xi \to \pm\pi/h} \hat{\mathbf{v}}^h_{3,ph}(\xi) = \left(\frac{\sqrt{3}}{\sqrt{7}} \quad \pm\frac{2}{i\sqrt{7}}\right)^*$.

*(a4)* $\lim_{\xi \to \pm\pi/h} i\,\hat{v}^h_{s,ph}(\xi) = \pm\infty$ *and* $\lim_{\xi \to \pm\pi/h} \hat{\mathbf{v}}^h_{s,ph}(\xi) = \left(0 \quad \pm\frac{1}{i}\right)^*$ *for all* $s \in (1, 3)$.

**Proposition 4.2.** *The spurious eigenvector $\hat{\mathbf{v}}^h_{s,sp}(\xi)$ has the following properties (see Fig. 4.2):*

*(b1)* $\lim_{\xi \to 0} \hat{v}^h_{s,sp}(\xi) = 0$ *and* $\lim_{\xi \to 0} \hat{\mathbf{v}}^h_{s,sp}(\xi) = (0 \quad 1)^*$ *for all* $s > 1$.

*(b2)* $\lim_{\xi \to \pm\pi/h} i\,\hat{v}^h_{s,sp}(\xi) = 0$ *and* $\lim_{\xi \to \pm\pi/h} \hat{\mathbf{v}}^h_{s,sp}(\xi) = (0 \quad 1)^*$ *for all* $s > 3$.

*(b3)* $\lim_{\xi \to \pm\pi/h} i\,\hat{v}^h_{3,sp}(\xi) = \pm\sqrt{3}/2$ *and* $\lim_{\xi \to \pm\pi/h} \hat{\mathbf{v}}^h_{3,sp}(\xi) = \left(\pm\frac{\sqrt{3}}{i\sqrt{7}} \quad \frac{2}{\sqrt{7}}\right)^*$.

*(b4)* $\lim_{\xi \to \pm\pi/h} i\,\hat{v}^h_{s,sp}(\xi) = \pm\infty$ *and* $\lim_{\xi \to \pm\pi/h} \hat{\mathbf{v}}^h_{s,sp}(\xi) = \left(\pm\frac{1}{i} \quad 0\right)^*$ *for all* $s \in (1, 3)$.

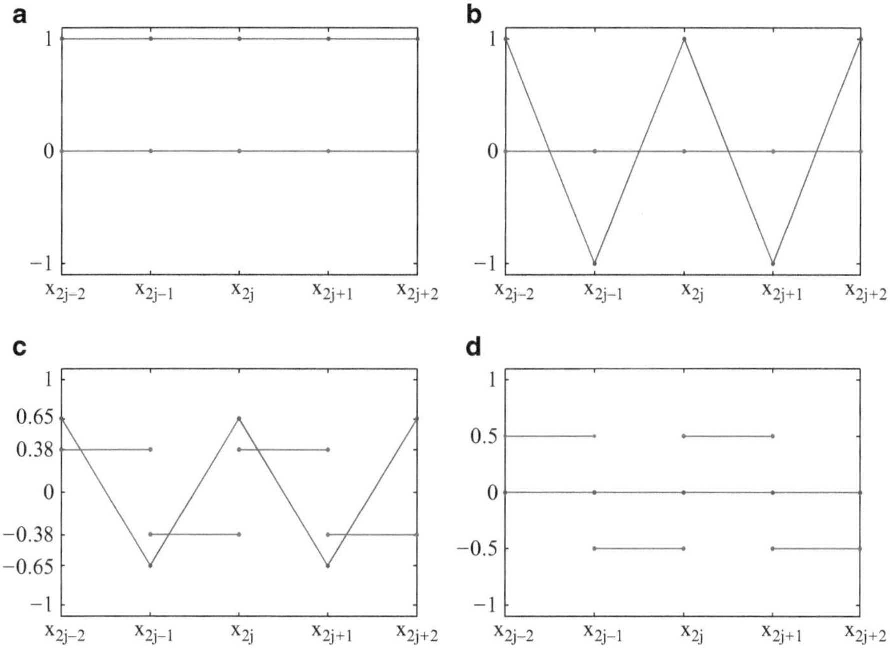


**Fig. 4.1** Plane waves for the SIPG method: real/imaginary part (*blue/red*) of the function $f^h \in \mathcal{V}^h$ whose coefficients with respect to the DG basis are $(\{f^h\}(x_j), [f^h](x_j))^* := \hat{\mathbf{v}}^h_{s,ph}(\xi_0)\exp(i\xi_0 x_j)$. The four subfigures correspond to: (**a**) $s > 1$, $\xi_0 = 0$ (**b**) $s > 3$, $\xi_0 = \pi/h$ (**c**) $s = 3$, $\xi_0 = \pi/h$ (**d**) $s \in (1,3)$, $\xi_0 = \pi/h$

*Proof of Propositions 4.1 and 4.2.* The conclusion of the two propositions follows by elementary calculus. More precisely, the discontinuity appearing in each eigenvector at $(s,\xi) = (3, \pi/h)$ is due to the fact that the following limit

$$\lim_{\xi\to\pm\pi/h}\left(2-\cos(\xi h)-\frac{12\left(s-\cos^2\left(\frac{\xi h}{2}\right)\right)}{h^2\hat{\Lambda}^h_{s,sp}(\xi)}\right)=\begin{cases}3-s, & s<3\\ 0, & s\geq 3\end{cases} \tag{4.9}$$

behaves differently according to the values of $s$. □

We denote by $\hat{\mathbf{V}}^h_s(\xi)$ the $2\times 2$-matrix whose columns are the eigenvectors $\hat{\mathbf{v}}^h_{s,ph}(\xi)$ and $\hat{\mathbf{v}}^h_{s,sp}(\xi)$. Namely,

$$\hat{\mathbf{V}}^h_s(\xi)=\begin{pmatrix}\frac{1}{\sqrt{1+|\hat{v}^h_{s,ph}(\xi)|^2}} & \frac{\hat{v}^h_{s,sp}(\xi)}{\sqrt{1+|\hat{v}^h_{s,sp}(\xi)|^2}}\\ \frac{\hat{v}^h_{s,ph}(\xi)}{\sqrt{1+|\hat{v}^h_{s,ph}(\xi)|^2}} & \frac{1}{\sqrt{1+|\hat{v}^h_{s,sp}(\xi)|^2}}\end{pmatrix}. \tag{4.10}$$

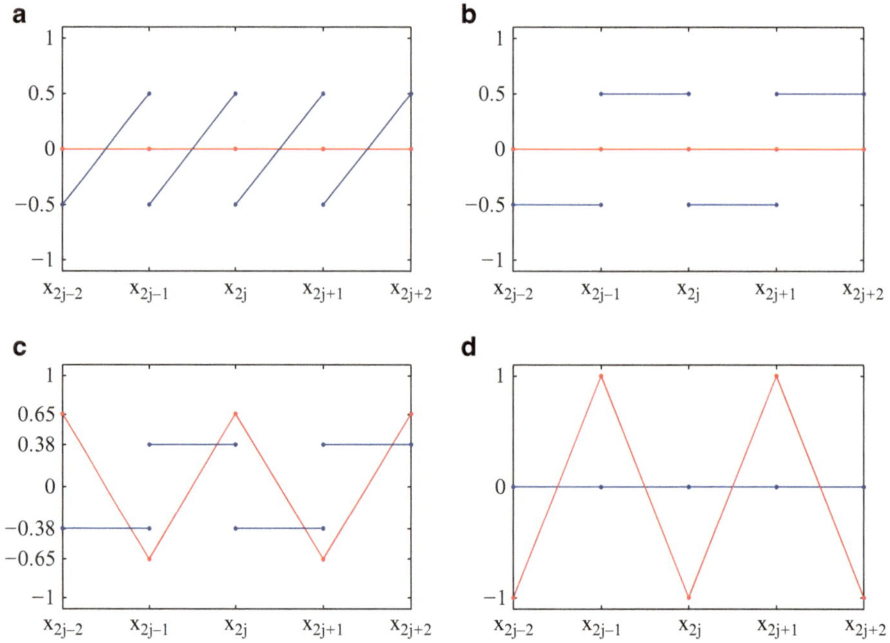


**Fig. 4.2** Plane waves for the SIPG method: real/imaginary part (*blue/red*) of the function $f^h \in \mathcal{V}^h$ whose coefficients with respect to the DG basis are $(\{f^h\}(x_j), [f^h](x_j))^* := \hat{\mathbf{v}}^h_{s,sp}(\xi_0)\exp(i\xi_0 x_j)$. The four subfigures correspond to: (**a**) $s > 1$, $\xi_0 = 0$ (**b**) $s > 3$, $\xi_0 = \pi/h$, (**c**) $s = 3$, $\xi_0 = \pi/h$ (**d**) $s \in (1,3)$, $\xi_0 = \pi/h$

The matrix $\hat{\mathbf{S}}^h_s(\xi)$ can be decomposed as follows:

$$\hat{\mathbf{S}}^h_s(\xi) = \hat{\mathbf{V}}^h_s(\xi)\begin{pmatrix} \hat{\Lambda}^h_{s,ph}(\xi) & 0 \\ 0 & \hat{\Lambda}^h_{s,sp}(\xi) \end{pmatrix}\left(\hat{\mathbf{V}}^h_s(\xi)\right)^{-1}, \tag{4.11}$$

so that the solution of (2.13) is given by

$$\begin{aligned} \hat{\mathbf{u}}^h_s(\xi,t) =& \frac{1}{2}\sum_{\pm}\hat{\mathbf{V}}^h_s(\xi)\begin{pmatrix} \exp(\pm it\hat{\lambda}^h_{s,ph}(\xi)) & 0 \\ 0 & \exp(\pm it\hat{\lambda}^h_{s,sp}(\xi)) \end{pmatrix}\left(\hat{\mathbf{V}}^h_s(\xi)\right)^{-1}\hat{\mathbf{u}}^{h,0}(\xi) \\ &+ \frac{1}{2}\sum_{\pm}\hat{\mathbf{V}}^h_s(\xi)\begin{pmatrix} \pm\frac{\exp(\pm it\hat{\lambda}^h_{s,ph}(\xi))}{i\hat{\lambda}^h_{s,ph}(\xi)} & 0 \\ 0 & \pm\frac{\exp(\pm it\hat{\lambda}^h_{s,sp}(\xi))}{i\hat{\lambda}^h_{s,sp}(\xi)} \end{pmatrix}\left(\hat{\mathbf{V}}^h_s(\xi)\right)^{-1}\hat{\mathbf{u}}^{h,1}(\xi). \end{aligned} \tag{4.12}$$

## 4.2 Analysis of the Two Group Velocities

For $s \in (1,\infty) \setminus \{3\}$, the *physical/spurious group velocities* take the following explicit form (see Fig. 4.3):

$$\partial_\xi \hat{\lambda}^h_{s,\alpha}(\xi) = -\mathrm{sign}(\alpha)\frac{2\sin\left(\frac{\xi h}{2}\right)}{\sqrt{h^2\hat{\Lambda}^h_{s,\alpha}(\xi)}}\frac{\cos\left(\frac{\xi h}{2}\right)}{\sqrt{\hat{\Delta}^h_s(\xi)}}\hat{e}^h_{s,\alpha}(\xi), \tag{4.13}$$

where $\alpha \in \{ph, sp\}$, $\mathrm{sign}(ph) = -1$, $\mathrm{sign}(sp) = 1$ and

$$\hat{e}^h_{s,\alpha}(\xi) = (s-3)h^2\hat{\Lambda}^h_{s,\alpha}(\xi) + 6(s - \cos(\xi h)). \tag{4.14}$$

When $s = 3$, the *physical/spurious group velocities* take the form

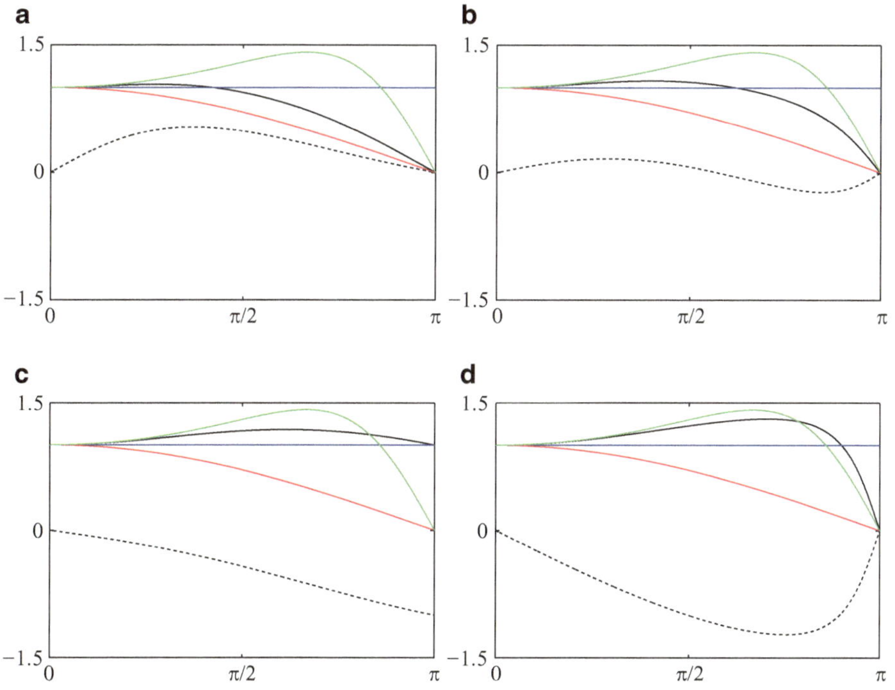


**Fig. 4.3** The physical/spurious group velocity for the SIPG method, $\partial_\xi \hat{\lambda}^1_{s,\alpha}(\xi)$, $\alpha = ph/sp$ (*black/dotted black*), compared to the ones of the continuous wave equation $\partial_\xi \hat{\lambda}(\xi) \equiv 1$ (*blue*) and of its FD and $P_1$-FEM semi-discretizations, $\partial_\xi \hat{\lambda}^1_{1,ph}(\xi)$ (*red*) and $\partial_\xi \hat{\lambda}^1_{\infty,ph}(\xi)$ (*green*). The four subfigures correspond to: **(a)** $s = 1.5$ **(b)** $s = 2$ **(c)** $s = 3$ **(c)** $s = 3$ **(d)** $s = 5$

$$\partial_\xi \hat{\lambda}^h_{3,ph}(\xi) = \sqrt{\frac{3 + \cos\left(\frac{\xi h}{2}\right)\sqrt{9 + 3\sin^2\left(\frac{\xi h}{2}\right)}}{3\left(2 + \sin^2\left(\frac{\xi h}{2}\right)\right)}} \frac{3\left(1 + \sin^2\left(\frac{\xi h}{2}\right)\right)}{\sqrt{9 + 3\sin^2\left(\frac{\xi h}{2}\right)}}$$

and

$$\partial_\xi \hat{\lambda}^h_{3,sp}(\xi) = -\frac{\sin\left(\frac{\xi h}{2}\right)}{\sqrt{3 + \cos\left(\frac{\xi h}{2}\right)\sqrt{9 + 3\sin^2\left(\frac{\xi h}{2}\right)}}} \frac{3\left(1 + \sin^2\left(\frac{\xi h}{2}\right)\right)}{\sqrt{9 + 3\sin^2\left(\frac{\xi h}{2}\right)}}.$$

The following two results characterize the behavior of the two group velocities.

**Proposition 4.3.** *The physical group velocity* $\partial_\xi \hat{\lambda}^h_{s,ph}(\xi)$ *has the following properties:*

*(c1) For all* $s > 1$, $\lim_{\xi \to 0} \partial_\xi \hat{\lambda}^h_{s,ph}(\xi) = 1$.

*(c2) For all* $s \in (1, \infty) \setminus \{3\}$, $\lim_{\xi \to \pm\pi/h} \partial_\xi \hat{\lambda}^h_{s,ph}(\xi) = 0$.

*(c3)* $\lim_{\xi \to \pm\pi/h} \partial_\xi \hat{\lambda}^h_{3,ph}(\xi) = 1$.

*(c4) For all* $s \in (1, \infty)$ *and all* $\xi \in \Pi^h$, $\hat{e}^h_{s,ph}(\xi) > 0$.

**Proposition 4.4.** *The spurious group velocity* $\partial_\xi \hat{\lambda}^h_{s,sp}(\xi)$ *has the following properties:*

*(d1) For all* $s > 1$, $\lim_{\xi \to 0} \partial_\xi \hat{\lambda}^h_{s,sp}(\xi) = 0$.

*(d2) For all* $s \in (1, \infty) \setminus \{3\}$, $\lim_{\xi \to \pm\pi/h} \partial_\xi \hat{\lambda}^h_{s,sp}(\xi) = 0$.

*(d3)* $\lim_{\xi \to \pm\pi/h} \partial_\xi \hat{\lambda}^h_{3,sp}(\xi) = -1$.

*(d4) For all* $s \in [5/2, \infty)$ *and all* $\xi \in \Pi^h$, $\hat{e}^h_{s,sp}(\xi) > 0$.

*(d5) For all* $s \in (1, 5/3]$ *and all* $\xi \in \Pi^h$, $\hat{e}^h_{s,sp}(\xi) < 0$.

*(d6) For all* $s \in (5/3, 5/2)$, $\hat{e}^h_{s,sp}(0) < 0$, $\hat{e}^h_{s,sp}(\pi/h) > 0$, *and there exists a unique wave number* $\xi_s \in (0, \pi/h)$ *such that* $\hat{e}^h_{s,sp}(\xi_s) = 0$, *whose description is given by*

$$\cos(\xi_s h) = \frac{s(2s-3) - (s-1)(3-s)\sqrt{6(s-1)}}{3 - (3-s)^2}. \tag{4.15}$$

*Moreover,* $\xi_{5/2} = 0$ *and* $\xi_{5/3} = \pi/h$.

The proofs of Propositions 4.3 and 4.4 are more technical and they will be given in Appendix A.

*Remark 4.2.* Propositions 4.3 and 4.4 provide information concerning the following quantitative and qualitative properties of the two dispersion diagrams:

- Property (c1) is related to the convergence as $h \to 0$ of the SIPG approximations to the continuous wave equation, for which the group velocity is identically one.
- Property (c2) is usual for the group velocities of some classical approximations of the wave equation on uniform meshes (e.g., FD or $P_1$-FEM). As a consequence of that, there are high-frequency wave packets concentrated on the physical dispersion diagram propagating at arbitrarily low speed.
- Property (c3) is related to the fact that, as proved in Lemma 4.1, $\hat{\Delta}_s^h(\xi)$ vanishes as $\xi \to \pm\pi/h$ for $s = 3$.
- Property (c4) means that the unique critical point of $\hat{\lambda}^h_{s,ph}(\xi)$ is $\xi = \pm\pi/h$. Thus, $\partial_\xi \hat{\lambda}^h_{s,ph}(\xi) > 0$ for all $\xi \in [0, \pi/h)$ and $s > 1$, so that $\hat{\lambda}^h_{s,ph}(\xi)$ is strictly increasing in $\xi$ (see Fig. 2.2). As a consequence of this monotonicity property of the physical dispersion relation, we obtain the following upper bound of $\hat{\lambda}^h_{s,ph}(\xi)$ in terms of $s$:

$$\max_{\xi \in \Pi^h} |\hat{\lambda}^h_{s,ph}(\xi)| = |\hat{\lambda}^h_{s,ph}(\pm\pi/h)| = \begin{cases} \frac{2\sqrt{s}}{h}, & s \in (1,3) \\ \frac{2\sqrt{3}}{h}, & s \geq 3. \end{cases}$$

- The spurious diagram $\hat{\lambda}^h_{s,sp}(\xi)$ exhibits a rich behavior from the point of view of its monotonicity, in several ranges of the penalization parameter $s$, that can be summarized as follows (see also Fig. 2.2):
  - For $s \geq 5/2$, $\hat{\lambda}^h_{s,sp}(\xi)$ is decreasing on $(0, \pi/h)$, with minimal value in $\xi = \pi/h$ equal to $2\sqrt{\max\{s,3\}}/h$ and the maximal one at $\xi = 0$, equal to $2\sqrt{3(s-1)}/h$.
  - For $s \in (1, 5/3)$, $\hat{\lambda}^h_{s,sp}(\xi)$ is strictly increasing on $(0, \pi/h)$, with maximal value in $\xi = \pi/h$ equal to $2\sqrt{3}/h$ and the minimal one at $\xi = 0$, equal to $2\sqrt{3(s-1)}/h$.
  - For $s \in (5/3, 5/2)$, $\hat{\lambda}^h_{s,sp}(\xi)$ has two minimum points located at $\xi = 0$ and $\xi = \pi/h$, where the value of the spurious dispersion relation is $2\sqrt{3(s-1)}/h$ and $2\sqrt{3}/h$, respectively. Moreover, it has a unique maximum point located at the wave number $\xi_s \in (0, \pi/h)$, whose precise description is given by (4.15). By plugging the expression of $\xi_s$ in (4.15) into the spurious dispersion relation, we obtain that its value at the maximum point is given by

$$\hat{\lambda}^h_{s,sp}(\xi_s) = \frac{1}{h}\sqrt{\frac{6(s-1)(s+\sqrt{6(s-1)})}{3-(3-s)^2}} \quad \forall s \in (5/3, 5/2).$$

# Chapter 5
# On the Lack of Uniform Observability for Discontinuous Galerkin Approximations of Waves

The purpose of this chapter is to construct initial data for the discontinuous Galerkin approximations of the wave equation so that the corresponding wave packets propagate arbitrarily slowly. Our construction is based on the fact that on each Fourier mode there are critical wave numbers at which the corresponding group velocity vanishes. Moreover, we prove that the observability constant blows up at least polynomially at any order as the mesh size $h$ tends to zero.

This justifies the need of designing efficient high-frequency filtering mechanisms.

Our goal being to build specific classes of pathological solutions, we consider initial data $(\hat{\mathbf{u}}^{h,0}(\xi), \hat{\mathbf{u}}^{h,1}(\xi))$ in (2.13) satisfying two requirements:

- The first one (which will be also used in the first two filtering algorithms within the next chapter) requires initial data *concentrated on the physical branch* of the dispersion relation, so that the following condition is fulfilled:

$$\hat{\mathbf{u}}^{h,i}(\xi) = \hat{\mathbf{v}}^h_{s,ph}(\xi)\hat{u}^{h,i}(\xi), \tag{5.1}$$

  where $i = 0, 1$, $\hat{\mathbf{v}}^h_{s,ph}(\xi)$ is the *physical eigenvector* introduced in (4.7) and $\hat{u}^{h,i}(\xi)$ are scalar functions defined on $\Pi^h := [-\pi/h, \pi/h]$.
- The second restriction we impose on the data is that $(\hat{u}^{h,0}(\xi), \hat{u}^{h,1}(\xi))$ in (5.1) are related as follows:

$$\hat{u}^{h,1}(\xi) = i\hat{\lambda}^h_{s,ph}(\xi)\hat{u}^{h,0}(\xi). \tag{5.2}$$

For all initial data satisfying property (5.1), the solution is *concentrated on the physical mode*. More precisely, since $(\hat{\mathbf{V}}^h_s(\xi))^{-1}\hat{\mathbf{v}}^h_{s,ph}(\xi) = (1 \quad 0)^*$, the solution of (2.13), given initially by (4.12), can be simply represented only in terms of the physical eigensolution as

$$\hat{\mathbf{u}}^h(\xi, t) = \frac{1}{2}\hat{\mathbf{v}}^h_{s,ph}(\xi)\sum_{\pm}\Big(\hat{u}^{h,0}(\xi) \pm \frac{\hat{u}^{h,1}(\xi)}{i\hat{\lambda}^h_{s,ph}(\xi)}\Big)\exp(\pm it\hat{\lambda}^h_{s,ph}(\xi)). \tag{5.3}$$

A. Marica and E. Zuazua, *Symmetric Discontinuous Galerkin Methods for 1 – D Waves*, SpringerBriefs in Mathematics, DOI 10.1007/978-1-4614-5811-1_5,

The associated *total energy* can also be simplified with respect to (2.15) as follows:

$$\mathscr{E}_s^{W,h}(\mathbf{u^{h,0}}, \mathbf{u^{h,1}}) = \frac{1}{4\pi} \int_{\Pi^h} \left( \hat{m}^h_{s,ph}(\xi) |\hat{u}^{h,1}(\xi)|^2 + \hat{r}^h_{s,ph}(\xi) |\hat{u}^{h,0}(\xi)|^2 \right) d\xi, \quad (5.4)$$

where

$$\hat{m}^h_{s,ph}(\xi) = (\hat{\mathbf{M}}^h(\xi)\hat{\mathbf{v}}^h_{s,ph}(\xi), \hat{\mathbf{v}}^h_{s,ph}(\xi))_{\mathbb{C}^2} \text{ and } \hat{r}^h_{s,ph}(\xi) = (\hat{\mathbf{R}}^h_s(\xi)\hat{\mathbf{v}}^h_{s,ph}(\xi), \hat{\mathbf{v}}^h_{s,ph}(\xi))_{\mathbb{C}^2}.$$

**Lemma 5.1.** *For all* $\xi \in \Pi^h$ *and all* $s \in (1, \infty)$, *the following identity holds:*

$$\hat{r}^h_{s,ph}(\xi) = \hat{\Lambda}^h_{s,ph}(\xi)\hat{m}^h_{s,ph}(\xi). \quad (5.5)$$

*Proof.* Since $\hat{\mathbf{v}}^h_{s,ph}(\xi)$ is the eigenvector of the matrix $\hat{\mathbf{S}}^h_s(\xi) = (\hat{\mathbf{M}}^h(\xi))^{-1}\hat{\mathbf{R}}^h_s(\xi)$ corresponding to the physical eigenvalue $\hat{\Lambda}^h_{s,ph}(\xi)$, we have the following identities:

$$\begin{aligned}\hat{r}^h_{s,ph}(\xi) &= (\hat{\mathbf{M}}^h(\xi)\hat{\mathbf{S}}^h_s(\xi)\hat{\mathbf{v}}^h_{s,ph}(\xi), \hat{\mathbf{v}}^h_{s,ph}(\xi))_{\mathbb{C}^2} = \hat{\Lambda}^h_{s,ph}(\xi)(\hat{\mathbf{M}}^h(\xi)\hat{\mathbf{v}}^h_{s,ph}(\xi), \hat{\mathbf{v}}^h_{s,ph}(\xi))_{\mathbb{C}^2} \\ &= \hat{\Lambda}^h_{s,ph}(\xi)\hat{m}^h_{s,ph}(\xi),\end{aligned}$$

which yield precisely (5.5). □

Taking Lemma 5.1 into account and in view of the second condition (5.2), the expression of the solution and its total energy are simplified with respect to (5.3) and (5.4), taking the form

$$\hat{\mathbf{u}}^h(\xi, t) = \hat{\mathbf{v}}^h_{s,ph}(\xi)\hat{u}^{h,0}(\xi)\exp(it\hat{\lambda}^h_{s,ph}(\xi)) \quad (5.6)$$

and

$$\mathscr{E}_s^{W,h}(\mathbf{u^{h,0}}, \mathbf{u^{h,1}}) = \frac{1}{2\pi} \int_{\Pi^h} \hat{r}^h_{s,ph}(\xi)|\hat{u}^{h,0}(\xi)|^2 \, d\xi. \quad (5.7)$$

Thus, by applying the inverse SDFT (cf. [37]) to the solution $\hat{\mathbf{u}}^h(\xi, t)$ in (5.7), we obtain that the solution of the SIPG approximation (2.6) for data satisfying (5.1) and (5.2) is a wave packet whose phase is precisely the physical dispersion relation $\hat{\lambda}^h_{s,ph}(\xi)$. According to the monotonicity of $\hat{\lambda}^h_{s,ph}(\xi)$, both jump and average parts of these solutions of (2.6) propagate only in the negative direction, along the physical rays of geometric optics $x^{h,-}_{s,ph}(t)$, where

$$x^{h,\pm}_{s,\alpha}(t) = x^* \pm \partial_\xi \hat{\lambda}^h_{s,\alpha}(\xi_0)t, \quad \alpha \in \{ph, sp\},$$

$x^* \in \mathbb{R}$ and $\xi_0 \in \Pi^h$. In particular, for $\xi_0$ close to $\pi/h$, due to property (c2) in Proposition 4.3 stating that the physical group velocity vanishes at $\xi = \pi/h$, the rays $x^{h,-}_{s,ph}(t)$ are almost vertical.

There is an extensive literature concerning the behavior of wave packets and, in particular, of those for which the phase is singular (see, e.g., [65] or [67]). In particular, we can obtain the following result similar to the ones for the FD or $P_1$-FEM approximations in [28] or [49], stating the blowup as $h \to 0$ of the observability constant $C_s^h(T)$ in (2.9) at least at an arbitrary polynomial rate. The corresponding proof is based on an adaption of the stationary phase lemma (cf. [65]).

**Theorem 5.1.** *Fix $T > 0$ and consider a wave number $\xi_0 \in \Pi^h \setminus \{0\}$ and a starting point $x^* \in (-1,1)$ for the ray $x_{s,ph}^{h,-}(t)$ such that*

$$|x_{s,ph}^{h,-}(t)| < 1, \ \forall t \in [0,T], \tag{5.8}$$

*so that $x_{s,ph}^{h,-}(t)$ does not enter the observation region $\Omega := \mathbb{R} \setminus (-1,1)$ before time $T$.*

*Consider $\hat{\sigma} \in C_c^\infty(-1,1)$ and $\gamma = \gamma(h) > 0$ to be a function such that the following two conditions are satisfied:*

$$\lim_{h\to 0} \gamma(h) = +\infty \text{ and } \lim_{h\to 0} h\gamma(h) = 0. \tag{5.9}$$

*We also consider initial data $(\mathbf{u}^{\mathbf{h},0}, \mathbf{u}^{\mathbf{h},1})$ in (2.6) satisfying the two requirements (5.1) and (5.2), with $\hat{u}^{h,0}(\xi)$ given by*

$$\hat{u}^{h,0}(\xi) = \gamma^{-1/2}\hat{\sigma}(\gamma^{-1}(\xi - \xi_0))(\hat{r}_{s,ph}^h(\xi))^{-1/2} \exp(-i\xi x^*). \tag{5.10}$$

*Then, for all $\theta > 0$, there exists a constant $C_{s,\theta} = C_{s,\theta}(T, \hat{\sigma}, \xi_0) > 0$, not depending on h, such that the discrete observability constant $C_s^h(T)$ in (2.9) satisfies $C_s^h(T) \geq C_{s,\theta}\gamma^\theta$.*

In Figs. 5.1 and 5.2, we plot $|u(x,t)|$, where $u(x,t)$ is the solution of the continuous wave equation (1.18) with initial position $u^0(x) = \exp(-\gamma x^2/2)\exp(i\xi_0 x)$, $\gamma = h^{-1/2}$, and initial velocity $u^1 = u_x^0$, so that $u(x,t) = u^0(x+t)$ and the continuous solution propagates to the left with unit speed. We also plot $(|\{u_s^h\}(x_j,t)|^2 + |[u_s^h](x_j,t)|^2)^{1/2}$, where $\mathbf{u}_s^{\mathbf{h}}(t)$ is the solution of the SIPG approximation (2.6) with $s = 5$, initial data as in Theorem 5.1, with $\sigma(x) = \exp(-x^2/2)$ and final time $T = 1$. We observe that

- For $\xi_0 = 9\pi/(10h)$, the numerical solution propagates at almost unit velocity, i.e., $\partial_\xi \hat{\lambda}_{5,ph}^1(9\pi/10) = 0.9953$.
- Nevertheless, for $\xi_0 = 49\pi/(50h)$ and $\xi_0 = 99\pi/(100h)$, it propagates at much lower group velocities, $\partial_\xi \hat{\lambda}_{5,ph}^1(49\pi/50) = 0.2675$ and $\partial_\xi \hat{\lambda}_{5,ph}^1(99\pi/100) = 0.1355$.

We also emphasize that the dispersion phenomena for the case where $\xi_0 \sim \pi/h$ (e.g., $\xi_0 = 99\pi/(100h)$ and $h = 1/2000$) are considerably reduced when $h = 1/5000$.

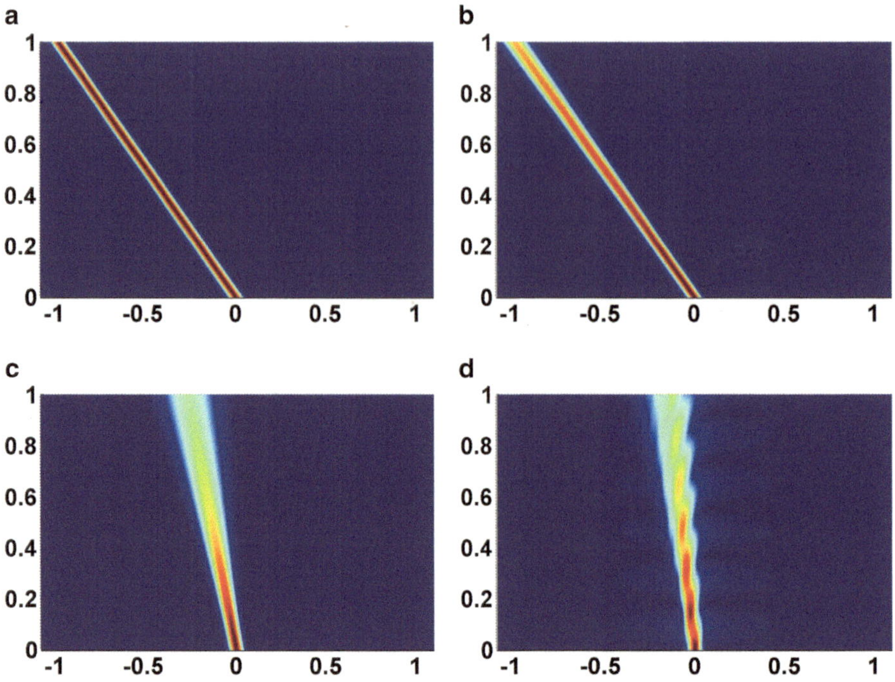


**Fig. 5.1** Gaussian solutions of the continuous wave equation versus the ones of its DG approximation in Theorem 5.1, concentrated on the physical mode, with $h = 1/2000$. The four subplots correspond to: (**a**) Continuous (**b**) $\eta_0 = 9\pi/10$ (**c**) $\eta_0 = 49\pi/50$ (**d**) $\eta_0 = 99\pi/100$

A similar result to Theorem 5.1 holds for solutions of the SIPG approximation concentrated on the spurious diagram. In that case, according to Proposition 4.4, the wave number $\xi_0$ determining the ray $x^{h,\pm}_{s,sp}(t)$ should be close to $\xi = 0$, to $\xi = \pi/h$ or to the third point of vanishing spurious group velocities in the case $s \in (5/3, 5/2)$, i.e., $\xi_s \in (0, \pi/h)$ in (4.15), in order to guarantee that the analogue of (5.8) is fulfilled.

In Figs. 5.3 and 5.4, we plot $(|\{u^h_s\}(x_j,t)|^2 + |[u^h_s](x_j,t)|^2)^{1/2}$, where $\mathbf{u}^{\mathbf{h}}_s(t)$ is the solution of the SIPG approximation (2.6) with $s = 5$, initial data as in Theorem 5.1, concentrated on the spurious mode, with $\sigma(x) = \exp(-x^2/2)$ and final time $T = 1$. Note that

- The direction of propagation which is to the right, opposite to the one for the physical mode in Figs. 5.1 and 5.2. This is due to the fact that, for $s = 5$, the physical/spurious group velocity is positive/negative for all $\xi \in [0, \pi/h]$ (see Fig. 4.3).

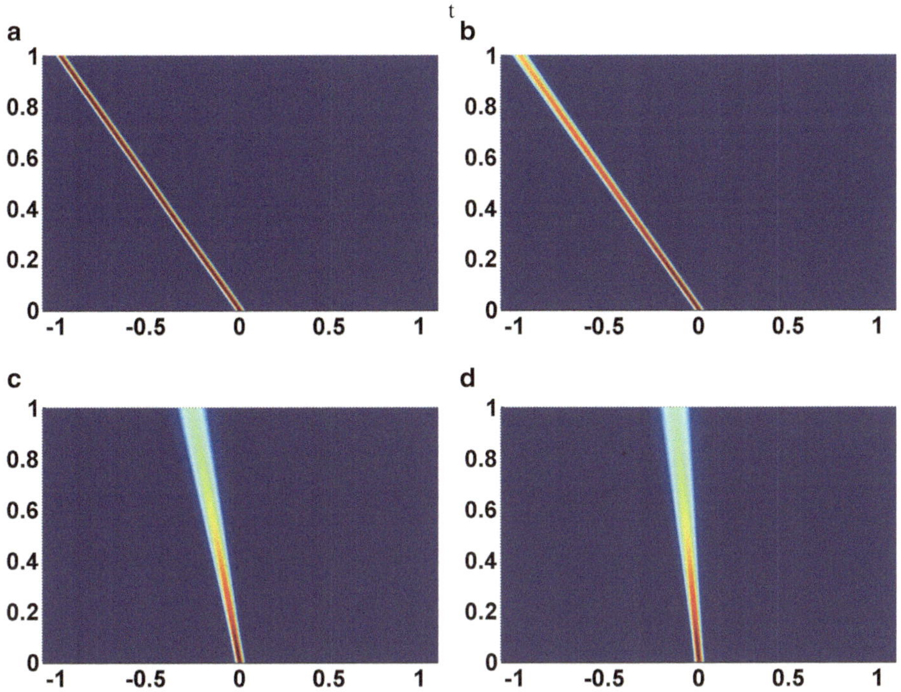


**Fig. 5.2** Gaussian solutions of the continuous wave equation versus the ones of its DG approximation in Theorem 5.1, concentrated on the physical mode, with $h = 1/5000$. The four subplots correspond to: (**a**) Continuous (**b**)$\eta_0 = 9\pi/10$ (**c**)$\eta_0 = 49\pi/50$ (**d**) $\eta_0 = 99\pi/100$

- When $\xi_0$ is in the low-frequency regime (e.g., $\xi_0 = \pi/100$, $\xi_0 = \pi/20$ or $\xi_0 = \pi/10$), the group velocity is small ($\partial_\xi \hat{\lambda}^1_{5,sp}(\pi/100) = -0.0227$, $\partial_\xi \hat{\lambda}^1_{5,sp}(\pi/50) = -0.0453$, $\partial_\xi \hat{\lambda}^1_{5,sp}(\pi/10) = -0.2257$), but the dispersion phenomena are almost absent.
- When $\xi_0 = 9\pi/10$, the velocity of propagation of the wave packet is $\partial_\xi \hat{\lambda}^1_{5,sp}(9\pi/10) = -0.9689$, while in the very high-frequency regime (i.e., $\xi_0 = 49\pi/(50h)$ and $\xi_0 = 99\pi/(100h)$), the speed of propagation is much smaller, i.e., $\partial_\xi \hat{\lambda}^1_{5,sp}(49\pi/50) = -0.2624$ and $\partial_\xi \hat{\lambda}^1_{5,sp}(99\pi/100) = -0.1329$.
- The dispersion phenomena are stronger for high-frequency wave packets than for the low-frequency ones for both values of $h$. This can be seen in the width of the support of the solution, increasing with the oscillation frequency $\eta_0$ as time evolves.

*Proof (of Theorem 5.1).* First we analyze the *total energy* of the solution of (2.6) corresponding to initial data $(\mathbf{u}^{\mathbf{h},0}, \mathbf{u}^{\mathbf{h},1})$ satisfying the two properties (5.1) and (5.10). Since $\gamma h \ll 1$, for $h$ small enough, $(\xi_0-\gamma, \xi_0+\gamma) \cap \Pi^h = (\xi_0-\gamma, \xi_0+\gamma)$ for all $\eta_0 \in \Pi^h$, $\xi_0 \neq \{\pm\pi/h\}$. Thus, using formula (5.7) for the total energy and

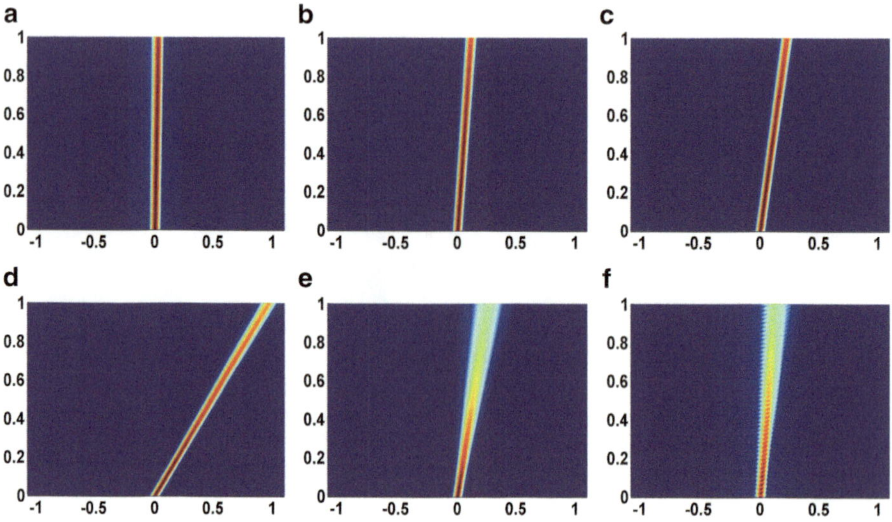


**Fig. 5.3** Gaussian solutions concentrated on the spurious mode of the DG approximation of waves with $h = 1/2000$. The six subplots correspond to: **(a)** $\eta_0 = \pi/100$ **(b)** $\eta_0 = \pi/20$ **(c)** $\eta_0 = \pi/10$ **(d)** $\eta_0 = 9\pi/10$ **(e )** $\eta_0 = 49\pi/50$ **(f)** $\eta_0 = 99\pi/100$

the fact that $\text{supp}(\hat{\sigma})(\gamma^{-1}(\cdot - \xi_0)) \cap \Pi^h = [\xi_0 - \gamma, \xi_0 + \gamma]$, we obtain that the total energy does not depend on $h$:

$$\begin{aligned}\mathscr{E}_s^{W,h}(\mathbf{u}^{\mathbf{h},0}, \mathbf{u}^{\mathbf{h},1}) &= \frac{1}{2\pi} \int_{\Pi^h} \gamma^{-1} |\hat{\sigma}(\gamma^{-1}(\xi - \xi_0))|^2 \, d\xi \\ &= \frac{1}{2\pi} \int_{\xi_0-\gamma}^{\xi_0+\gamma} \gamma^{-1} |\hat{\sigma}(\gamma^{-1}(\xi - \xi_0))|^2 \, d\xi = \frac{1}{2\pi} \int_{-1}^{1} |\hat{\sigma}(\eta)|^2 \, d\eta.\end{aligned} \tag{5.11}$$

For $\delta > 0$ and $t \in [0, T]$, set $\Omega_\delta(t) := \{x \in \mathbb{R} \text{ s.t. } |x - x_{s,ph}^{h,-}(t)| > \delta\}$. The following result holds:

**Lemma 5.2.** *Under the hypotheses of Theorem 5.1, for all $\delta > 0$, $N \in \mathbb{N}^*$, and all initial data $(\mathbf{u}^{\mathbf{h},0}, \mathbf{u}^{\mathbf{h},1})$ in (2.6), there exists a constant $C(N, T, \hat{\sigma}, \delta, \eta_0) > 0$, independent of h, such that*

$$\mathscr{E}_{s,\Omega_\delta(t)}^{W,h}(\mathbf{u}^{\mathbf{h}}(t), \partial_t \mathbf{u}^{\mathbf{h}}(t)) \le C(N, T, \hat{\sigma}, \delta, \eta_0) \gamma^{-(2N-1)} \quad \forall t \in [0, T]. \tag{5.12}$$

The conclusion of Theorem 5.1 follows since, if $T > 0$ and $\xi_0 \in \Pi^h$ are such that (5.8) is fulfilled, then there exists $\delta > 0$ such that $\{x \in \mathbb{R} \text{ s.t. } |x - x_{s,ph}^{h,-}(t)| < \delta\} \subset (-1, 1)$ for all $t \in [0, T]$. Thus, $\Omega = \mathbb{R} \setminus (-1, 1) \subset \bigcup_{t \in [0,T]} \Omega_\delta(t)$ and

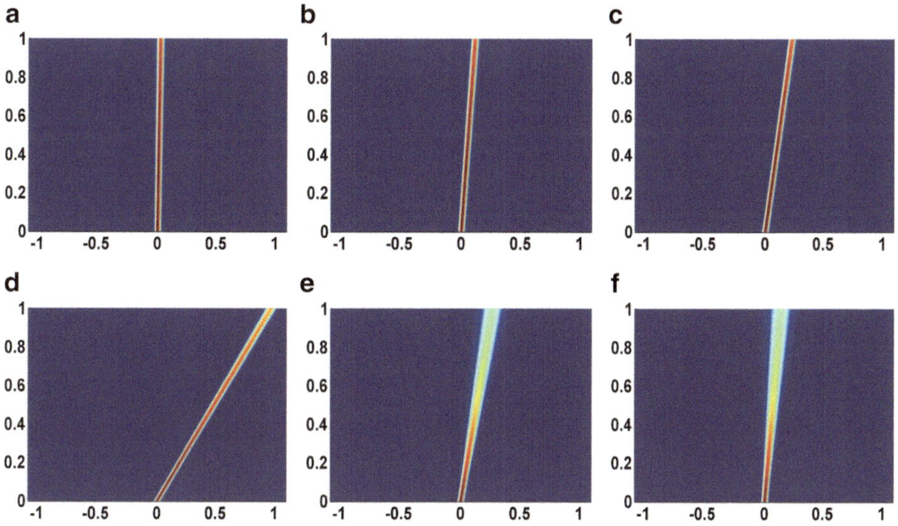


**Fig. 5.4** Gaussian solutions concentrated on the spurious mode of the DG approximation of waves with $h = 1/5000$. The six subplots correspond to: (**a**) $\eta_0 = \pi/100$ (**b**) $\eta_0 = \pi/20$ (**c**) $\eta_0 = \pi/10$ (**d**) $\eta_0 = 9\pi/10$ (**e**) $\eta_0 = 49\pi/50$ (**f**) $\eta_0 = 99\pi/100$

$$\int_0^T \mathscr{E}^{W,h}_{s,\Omega}(\mathbf{u^h}(t), \partial_t \mathbf{u^h}(t))\, dt \le \int_0^T \mathscr{E}^{W,h}_{s,\Omega_\delta(t)}(\mathbf{u^h}(t), \partial_t \mathbf{u^h}(t))\, dt$$
$$\le C(N, T, \hat{\sigma}, \delta, \eta_0) T \gamma^{-(2N-1)}. \qquad \square$$

*Proof (of Lemma 5.2).* Taking the following ingredients into account:

- The form (5.6) of the solution $\mathbf{u^h}(t)$ of (2.6) corresponding to initial data $(\mathbf{u^{h,0}}, \mathbf{u^{h,1}})$ concentrated on the physical mode (i.e., satisfying (5.1)):
- $\text{supp}(\hat{\sigma})(\gamma^{-1}(\cdot - \xi_0)) \cap \Pi^h = [\xi_0 - \gamma, \xi_0 + \gamma]$;
- $\hat{\mathbf{v}}^h_{s,ph}(\xi) = \hat{\mathbf{v}}^1_{s,ph}(\xi h)$;
- $\hat{\lambda}^h_{s,ph}(\xi) = \hat{\lambda}^1_{s,ph}(\xi h)/h$;
- $\hat{r}^h_{s,ph}(\xi) = \hat{r}^1_{s,ph}(\xi h)/h^2$,

we obtain the following explicit form of the two components of the solution with initial data given by (5.10):

$$\begin{pmatrix} \{u^h_s\}(x_j, t) \\ [u^h_s](x_j, t) \end{pmatrix} = \frac{\gamma^{1/2} h}{2\pi} \int_{-1}^{1} \frac{\hat{\sigma}(\eta) \hat{\mathbf{v}}^1_{s,ph}(\eta_0 + \gamma h \eta)}{\sqrt{\hat{r}^1_{s,ph}(\eta_0 + \gamma h \eta)}} \exp\left(\frac{i}{h} \psi(\gamma h \eta, x_j - x^*, t)\right) d\eta, \tag{5.13}$$

where the phase is of the form

$$\psi(\eta, x, t) := t\hat{\lambda}^1_{s,ph}(\eta_0 + \eta) + (\eta_0 + \eta)x.$$

In order to analyze the energy concentrated in $\Omega_\delta(t)$, we decompose it as follows (for simplicity of notation, put $I = \Omega_\delta(t)$):

$$\mathscr{E}^{W,h}_{s,I}(\mathbf{u}^{\mathbf{h}}_s(t), \partial_t \mathbf{u}^{\mathbf{h}}_s(t)) = \sum_{j=1}^{7} \mathscr{E}^j(\mathbf{u}^{\mathbf{h}}_s(t)), \tag{5.14}$$

where

$$\mathscr{E}^1(\mathbf{u}^{\mathbf{h}}_s(t)) := \frac{h}{6}\Big(||\mathscr{M}^{h,+}\{\partial_t \mathbf{u}^h_s(\cdot,t)\}||^2_{\ell^2(\mathscr{G}^h\cap I)} + ||\mathscr{M}^{h,-}\{\partial_t \mathbf{u}^h_s(\cdot,t)\}||^2_{\ell^2(\mathscr{G}^h\cap I)}\Big),$$

$$\mathscr{E}^2(\mathbf{u}^{\mathbf{h}}_s(t)) := \frac{h}{4}\Big(||\mathscr{D}^{h,+}\{\mathbf{u}^h_s(\cdot,t)\}||^2_{\ell^2(\mathscr{G}^h\cap I)} + ||\mathscr{D}^{h,-}\{\mathbf{u}^h_s(\cdot,t)\}||^2_{\ell^2(\mathscr{G}^h\cap I)}\Big),$$

$$\mathscr{E}^3(\mathbf{u}^{\mathbf{h}}_s(t)) := \frac{h^3}{96}\Big(||\mathscr{D}^{h,+}[\partial_t \mathbf{u}^h_s(\cdot,t)]||^2_{\ell^2(\mathscr{G}^h\cap I)} + ||\mathscr{D}^{h,-}[\partial_t \mathbf{u}^h_s(\cdot,t)]||^2_{\ell^2(\mathscr{G}^h\cap I)}\Big),$$

$$\mathscr{E}^4(\mathbf{u}^{\mathbf{h}}_s(t)) := \frac{h}{16}\Big(||\mathscr{D}^{h,+}[\mathbf{u}^h_s(\cdot,t)]||^2_{\ell^2(\mathscr{G}^h\cap I)} + ||\mathscr{D}^{h,-}[\mathbf{u}^h_s(\cdot,t)]||^2_{\ell^2(\mathscr{G}^h\cap I)}\Big),$$

$$\mathscr{E}^5(\mathbf{u}^{\mathbf{h}}_s(t)) = \frac{s-1}{2h}||[\mathbf{u}^h_s(\cdot,t)]||^2_{\ell^2(\mathscr{G}^h\cap I)}$$

and

$$\begin{aligned}\mathscr{E}^{(13\mp1)/2}(\mathbf{u}^{\mathbf{h}}_s(t)) :=& \frac{h}{48}||2\mathscr{M}^{h,\mp}\{\partial_t \mathbf{u}^h_s(\cdot,t)\} \pm \mathscr{M}^{h,\pm}[\partial_t \mathbf{u}^h_s(\cdot,t)]||^2_{\ell^2(\mathscr{G}^h\cap I)}\\ &+ \frac{h^3}{48}||\mathscr{D}^{h,\mp}\{\partial_t \mathbf{u}^h_s(\cdot,t)\} \pm \frac{1}{2}\mathscr{D}^{h,\pm}[\partial_t \mathbf{u}^h_s(\cdot,t)]||^2_{\ell^2(\mathscr{G}^h\cap I)}.\end{aligned}$$

Here $\mathscr{M}^{h,\pm}, \mathscr{D}^{h,\pm} : \ell^2(\mathscr{G}^h) \to \ell^2(\mathscr{G}^h)$ are the *forward* (+) and *backward* (-) discrete *mean* and *derivative* operators, defined as $\mathscr{M}^{h,\pm} f_j := (f_{j\pm1} + f_j)/2$ and $\mathscr{D}^{h,\pm} f_j := \pm(f_{j\pm1} - f_j)/h$.

Let us prove that the first term $\mathscr{E}^1(\mathbf{u}^{\mathbf{h}}(t))$ is polynomially small with respect to $h$, for the other six ones the arguments being similar. By taking one time derivative in the first component of (5.13) and applying the mean operator $\mathscr{M}^{h,\pm}$, we get

$$\mathscr{M}^{h,\pm}\{\partial_t u^h_s\}(x_j,t) = \frac{\gamma^{1/2}}{2\pi}\int_{-1}^{1} \widehat{\sigma}(\eta)\widehat{\theta}^{\pm}(\gamma h\eta) \exp\left(\frac{i}{h}\psi(\gamma h\eta, x_j - x^*, t)\right) d\eta, \tag{5.15}$$

where $\hat{v}^1_{s,ph}$ is as in (4.8) and

$$\widehat{\theta}^{\pm}(\eta) := \frac{i\widehat{\lambda}^1_{s,ph}(\eta_0+\eta)}{\sqrt{\widehat{r}^1_{s,ph}(\eta_0+\eta)}} \frac{\cos(\eta/2)\exp(\pm i\eta/2)}{\sqrt{1+|\widehat{v}^1_{s,ph}(\eta_0+\eta)|^2}}.$$

One can prove that $\hat{r}^1_{s,ph}$ in (5.4) is of the form $\hat{r}^1_{s,ph}(\eta) = 4\sin^2(\eta/2)\tilde{r}^1_{s,ph}(\eta)$, while $\hat{\lambda}^1_{s,ph}(\eta) = 2\sin(\eta/2)\tilde{\lambda}^1_{s,ph}(\eta)$, where $\tilde{r}^1_{s,ph}(\eta)$ and $\tilde{\lambda}^1_{s,ph}(\eta)$ are $C^\infty(\mathbb{R})$ functions on $[0,\pi]$. Moreover $\tilde{r}^1_{s,ph}(\eta) \geq 1$ for all $\eta \in [0,\pi]$. This means that $\hat{\lambda}^1_{s,ph}/\sqrt{\hat{r}^1_{s,ph}} \in C^\infty(0,\pi)$ and $\theta^\pm \in C^\infty(-1,1)$.

The complex exponential $\exp(i\psi(\gamma h\eta, x_j - x^*, t)/h)$ in (5.15) satisfies the following identity:

$$\exp\left(\frac{i}{h}\psi(\gamma h\eta, x_j - x^*, t)\right) = \frac{\partial_\eta\left(\exp\left(\frac{i}{h}\psi(\gamma h\eta, x_j - x^*, t)\right)\right)}{i\gamma\psi_\eta(\gamma h\eta, x_j - x^*, t)}. \tag{5.16}$$

We define the operator $\mathscr{L}$ as follows:

$$\mathscr{L}\hat{\sigma}(\eta, x, t) := \partial_\eta\left(\frac{1}{\psi_\eta(\gamma h\eta, x, t)}\hat{\sigma}(\eta)\hat{\theta}^\pm(\gamma h\eta)\right).$$

We apply the *stationary phase argument* (cf. [65]), integrating by parts in (5.15) $N$ times, applying iteratively the identity (5.16), and using the fact that $\hat{\sigma}\hat{\theta}^\pm(\gamma h\cdot) \in C_c^\infty(-1,1)$, so that the boundary terms vanish. Then

$$\begin{aligned}\mathscr{M}^{h,\pm}\{\partial_t u_s^h\}(x_j,t) &= \\ &= \frac{\gamma^{1/2}}{2\pi}\int_{-1}^{1}\left(-\frac{1}{i\gamma}\right)^N \mathscr{L}^N\widehat{\sigma}(\eta, x_j - x^*, t)\exp\left(\frac{i}{h}\psi(\gamma h\eta, x_j - x^*, t)\right)d\eta,\end{aligned} \tag{5.17}$$

so that, for all $j \in \mathbb{Z}$ and $t \in [0,T]$, we get

$$|\mathscr{M}^{h,\pm}\{\partial_t u_s^h\}(x_j,t)|^2 \leq \frac{2\gamma^{1-2N}}{4\pi^2}\int_{-1}^{1}|\mathscr{L}^N\widehat{\sigma}(\eta, x_j - x^*, t)|^2\,d\eta. \tag{5.18}$$

The second step of the argument based on the *stationary phase lemma* consists in using the fact that our aim is to estimate the energy for those $x_j$ which are *far from the ray* $x^{h,-}_{s,ph}(t)$. More precisely, for all $N \in \mathbb{N}^*$, we can prove that there exists a constant $C_N$ depending on $T$, $\delta$, $||\partial^\alpha\hat{\lambda}^1_{s,ph}||_{L^\infty(\eta_0-\gamma h,\eta_0+\gamma h)} \sim |\partial^\alpha\hat{\lambda}^1_{s,ph}(\eta_0)|$ for all $\alpha \leq N+1$ and $||\partial^\alpha\hat{\sigma}||_{L^\infty(-1,1)}$ for all $\alpha \leq N$ such that

$$|\mathscr{L}^N\hat{\sigma}(\eta, x_j - x^*, t)|^2 \leq \frac{C_N}{|x_j - x^{h,-}_{s,ph}(t)|^{2N}}, \tag{5.19}$$

for all $x_j \in \Omega_\delta(t)$ and for all $t \in [0,T]$.

Let us prove (5.19) for $N = 1$. For $N \geq 2$, the arguments are similar, but more technical. For $N = 1$, explicit computations give

$$\mathscr{L}\hat{\sigma}(\eta, x - x^*, t) = -\frac{\gamma h \partial_\eta^2 \psi(\eta_0 + \gamma h \eta, x - x^*, t)}{|\partial_\eta \psi(\eta_0 + \gamma h \eta, x - x^*, t)|^2} \hat{\sigma}(\eta) \hat{\theta}^{\pm}(\gamma h \eta) + \frac{\hat{\sigma}'(\eta)\hat{\theta}^{\pm}(\gamma h \eta) + \gamma h \hat{\sigma}(\eta)(\hat{\theta}^{\pm})'(\gamma h \eta)}{\partial_\eta \psi(\eta_0 + \gamma h \eta, x - x^*, t)}.$$

Observe that

$$\begin{aligned}\partial_\eta \psi(\eta_0 + \gamma h \eta, x - x^*, t) &= t \partial_\xi \hat{\lambda}^1_{s,ph}(\eta_0 + \gamma h \eta) + x - x^* \\ &= x - x^{h,-}_{s,ph}(t) + t \gamma h \eta \partial_\xi^2 \hat{\lambda}^1_{s,ph}(\eta_0 + \gamma h \eta'),\end{aligned}$$

with $\eta' \in (-1, 1)$. Then

$$|\partial_\eta \psi(\eta_0 + \gamma h \eta, x - x^*, t)| \geq |x - x^{h,-}_{s,ph}(t)| - T \gamma h ||\partial_\xi^2 \hat{\lambda}^1_{s,ph}||_{L^\infty(\eta_0 - \gamma h, \eta_0 + \gamma h)}.$$

Since $\gamma h \ll 1$, by (5.9) and the fact that the time $T$ is finite, we obtain the following inequalities:

$$T \gamma h ||\partial_\xi^2 \hat{\lambda}^1_{s,ph}||_{L^\infty(\eta_0 - \gamma h, \eta_0 + \gamma h)} \leq \frac{\delta}{2} \leq \frac{|x - x^{h,-}_{s,ph}(t)|}{2},$$

so that $|\partial_\eta \psi(\eta_0 + \gamma h \eta, x - x^*, t)| \geq |x - x^{h,-}_{s,ph}(t)|/2$.

On the other hand, $\partial_\eta^\alpha \psi(\eta_0 + \gamma h \eta, x - x^*, t) = t \partial_\eta^\alpha \hat{\lambda}^1_{s,ph}(\eta_0 + \gamma h \eta)$ does not depend anymore on $x$ for all $\alpha \geq 2$. This concludes (5.19) for $N = 1$.

From (5.18) and (5.19) we get

$$\mathscr{E}^1(\mathbf{u^h}(t)) = \frac{h}{6} \sum_{x_j \in \Omega_\delta(t)} |\mathscr{M}^{h,\pm}\{\partial_t u^h_s\}(x_j, t)|^2 \leq \frac{\gamma^{1-2N} C_N A}{6\pi^2} = O(\gamma^{1-2N}),$$

since, for all $\tilde{x} \in \mathbb{R}$ (and, in particular, for $\tilde{x} = x^{h,-}_{s,ph}(t)$), we get the following convergence of the Riemann sum as $h \to 0$:

$$A = h \sum_{|x_j - \tilde{x}| > \delta} |x_j - \tilde{x}|^{-2N} \to \int_{|x - \tilde{x}| > \delta} |x - \tilde{x}|^{-2N} \, dy = \int_{|y| > \delta} |y|^{-2N} \, dy = \frac{2\delta^{-(2N-1)}}{2N - 1}. \quad \square$$

# Chapter 6
# Filtering Mechanisms

In this chapter, we design several filtering strategies for the discontinuous Galerkin approximations of the wave and Klein–Gordon equations based on the classical *Fourier truncation* method or on the *bi-grid* filtering algorithm. We rigorously prove their efficiency to recover the uniform observability estimates as the mesh size parameter goes to zero. In the last section of this chapter, we present several numerical simulations showing, in particular, how solutions corresponding to bi-grid projections of highly oscillatory Gaussian profiles split into several wave packets.

## 6.1 Fourier Truncation on the Physical Mode

Before proceeding with the statement of the main result of this section, we introduce some useful notation. For $\delta \in (0,1)$, set $\Pi_\delta^h := [-\pi\delta/h, \pi\delta/h]$ and define the space of Fourier truncated data:

$$\mathcal{I}_\delta^h = \{\mathbf{f}^h \in \ell^2(\mathcal{G}^h) : \operatorname{supp}(\hat{f}^h) \subset \Pi_\delta^h\}. \tag{6.1}$$

We say that the initial data $\mathbf{u}^{h,i}, i = 0, 1$, whose SDFT is $\hat{u}^{h,i}(\xi)$ in (5.1), is obtained by the *Fourier truncation method* of parameter $\delta \in (0,1)$ if

$$\mathbf{u}^{h,i} \in \mathcal{I}_\delta^h \quad \forall i = 0, 1. \tag{6.2}$$

The following uniform observability result holds:

**Theorem 6.1.** *Consider initial data in (2.6) satisfying the two requirements (5.1) and (6.2). Then, for all $s > 1$ and $T > T^\star_{s,\mathrm{ph},\delta}$, the observability inequality (2.9) holds uniformly as $h \to 0$. Moreover, there exists $C_{s,\mathrm{ph},\delta} \in (0,1)$ such that the observability time $T^\star_{s,\mathrm{ph},\delta}$ is given by*

$$T^\star_{s,\mathrm{ph},\delta} = \frac{2}{v^\star_{s,\mathrm{ph},\delta}}(1 + C_{s,ph,\delta}), \tag{6.3}$$

A. Marica and E. Zuazua, *Symmetric Discontinuous Galerkin Methods for 1 – D Waves*, SpringerBriefs in Mathematics, DOI 10.1007/978-1-4614-5811-1_6,

*where* $v^{\star}_{s,\mathrm{ph},\delta}$ *is the minimal group velocity given below:*

$$v^{\star}_{s,\mathrm{ph},\delta} := \min_{\eta \in \Pi^h_\delta} \partial_\eta \hat{\lambda}^h_{s,\mathrm{ph}}(\eta). \tag{6.4}$$

For a better understanding of our methodology of proof of Theorem 6.1, let us first present its main steps in the context of the continuous wave equation, inspired by a classical proof for the dispersive estimates for the Schrödinger equation (cf. [43]).

**Step A.** From the time conservation of the total energy, for all $T > 0$ and all initial data $(u^0, u^1) \in \dot{H}^1(\mathbb{R}) \times L^2(\mathbb{R})$ in the continuous wave Eq. (1.18), we obtain the following identity relating the energies concentrated in $\Omega := \mathbb{R} \setminus (-1, 1)$ and $I := (-1, 1)$, with the notation of Sect. 1.3:

$$\int_0^T \mathscr{E}^W_\Omega(u(\cdot,t), \partial_t u(\cdot,t))\, dt = T\mathscr{E}^W(u^0, u^1) - \int_0^T \mathscr{E}^W_I(u(\cdot,t), \partial_t u(\cdot,t))\, dt. \tag{6.5}$$

**Step B.** We have the following obvious upper bound for the second term in the right-hand side of (6.5):

$$\int_0^T \mathscr{E}^W_I(u(\cdot,t), \partial_t u(\cdot,t))\, dt \le \int_{\mathbb{R}} \mathscr{E}^W_I(u(\cdot,t), \partial_t u(\cdot,t))\, dt. \tag{6.6}$$

**Step C.** We then prove

$$\int_{\mathbb{R}} \mathscr{E}^W_I(u(\cdot,t), \partial_t u(\cdot,t))\, dt \le 2\mathscr{E}^W(u^0, u^1). \tag{6.7}$$

Combining (6.5) and (6.7), we see that, in order to guarantee the positivity of the right-hand side in (6.5), we have to impose $T > 2$, obtaining the characteristic time $T^\star = 2$. In order to prove (6.7), we use the methodology in [43]. Let $\mathscr{F}_{t\to\tau}$ and $\mathscr{F}_{x\to\xi}$ be the *continuous Fourier transforms* with respect to the time/space variables $t$ and $x$.

**Step C1.** By the *Parseval identity*, we have

$$\int_{\mathbb{R}} \mathscr{E}^W_I(u(\cdot,t), \partial_t u(\cdot,t))\, dt$$
$$= \frac{1}{4\pi} \int_I \int_{\mathbb{R}} (|(\mathscr{F}_{t\to\tau} u_x)(x,\tau)|^2 + |(\mathscr{F}_{t\to\tau} u_t)(x,\tau)|^2)\, d\tau\, dx. \tag{6.8}$$

**Step C2.** Using the change of variable $\xi \to -\xi$ in the term corresponding to the $-$ sign in the Fourier representation of the continuous solution,

$$u(x,t) = \sum_{\pm} \frac{1}{4\pi} \int_{\mathbb{R}} \Big( (\mathscr{F}_{x\to\xi} u^0)(\xi) \pm \frac{1}{i\xi} (\mathscr{F}_{x\to\xi} u^1)(\xi) \Big) \exp(\pm i\xi t) \exp(i\xi x)\, d\xi, \tag{6.9}$$

we obtain the following expressions of the Fourier transforms in time in (6.8)

$$(\mathscr{F}_{t\to\tau}u_x)(x,\tau) = \sum_{\pm} \hat{u}^{\pm}(\tau)\exp(\pm i\tau x), \quad (\mathscr{F}_{t\to\tau}u_t)(x,\tau) = \sum_{\pm} \pm\hat{u}^{\pm}(\tau)\exp(\pm i\tau x), \tag{6.10}$$

where

$$\hat{u}^{\pm}(\tau) := \frac{1}{2}(i\tau(\mathscr{F}_{x\to\xi}u^0)(\pm\tau) + (\mathscr{F}_{x\to\xi}u^1)(\pm\tau)).$$

**Step C3.** The expression under the integral in the right-hand side of (6.8) does not depend on $x$. Indeed,

$$|(\mathscr{F}_{t\to\tau}u_x)(x,\tau)|^2 + |(\mathscr{F}_{t\to\tau}u_t)(x,\tau)|^2 = 2(|\hat{u}^+(\tau)|^2 + |\hat{u}^-(\tau)|^2). \tag{6.11}$$

**Step C4.** By the change of variable $\tau \to -\tau$ in the integral involving $\hat{u}^-(\tau)$, we obtain

$$\frac{1}{2\pi}\int_{\mathbb{R}}(|\hat{u}^+(\tau)|^2 + |\hat{u}^-(\tau)|^2)d\tau = \frac{1}{4\pi}\int_{\mathbb{R}}(\tau^2|(\mathscr{F}_{x\to\xi}u^0)(\tau)|^2 + |(\mathscr{F}_{x\to\xi}u^1)(\tau)|^2)\, d\tau. \tag{6.12}$$

Observe that the right-hand side in (6.12) is precisely the Fourier representation of the total energy $\mathscr{E}^W(u^0,u^1)$. Finally, inequality (6.12) combined with **Step C3** and the fact that $|I| = 2$ leads to (6.7).

*Proof (of Theorem 6.1).* Due to the time conservation of the total energy for the solutions of the SIPG approximation (2.6), we obtain the following discrete version of the identity (6.5) in **Step A**:

$$\int_0^T \mathscr{E}^{W,h}_{s,\Omega}(\mathbf{u}^{\mathbf{h}}_s(t), \partial_t\mathbf{u}^{\mathbf{h}}_s(t))\, dt = T\mathscr{E}^{W,h}_s(\mathbf{u}^{\mathbf{h},0}, \mathbf{u}^{\mathbf{h},1}) - \int_0^T \mathscr{E}^{W,h}_{s,I}(\mathbf{u}^{\mathbf{h}}_s(t), \partial_t\mathbf{u}^{\mathbf{h}}_s(t))\, dt. \tag{6.13}$$

We can also find an upper bound for the second term in the right-hand side of (6.13) of the form

$$\int_0^T \mathscr{E}^{W,h}_{s,I}(\mathbf{u}^{\mathbf{h}}_s(t), \partial_t\mathbf{u}^{\mathbf{h}}_s(t))\, dt \le \int_{\mathbb{R}} \mathscr{E}^{W,h}_{s,I}(\mathbf{u}^{\mathbf{h}}_s(t), \partial_t\mathbf{u}^{\mathbf{h}}_s(t))\, dt. \tag{6.14}$$

In order to proceed with **Step C** above, which requires to apply the Parseval identity, we split the energy concentrated in $I$ as in (5.14).

In what follows, we compute explicitly each one of the integrals

$$\mathscr{I}^j := \int_{\mathbb{R}} \mathscr{E}^j(\mathbf{u}^{\mathbf{h}}_s(t))\, dt \quad \forall 1 \le j \le 7. \tag{6.15}$$

As we did in **Step C1**, first we apply the Parseval identity in time, so that

$$\int_{\mathbb{R}} \mathscr{E}^j(\mathbf{u}_s^{\mathbf{h}}(t))\,dt = \frac{1}{2\pi}\int_{\mathbb{R}} \mathscr{E}^j(\mathscr{F}_{t\to\tau}(\mathbf{u}_s^{\mathbf{h}})(\tau))\,d\tau \quad \forall 1 \le j \le 7. \tag{6.16}$$

Under the hypothesis (5.1), the solution of (2.13) takes the form (5.3), so that, by the inverse SDFT, we obtain that

$$\begin{pmatrix} \{u_s^h\}(x_j,t) \\ [u_s^h](x_j,t) \end{pmatrix} = \frac{1}{2\pi}\sum_{\pm}\int_{\Pi_\delta^h} \frac{\hat{\mathbf{v}}_{s,\mathrm{ph}}^h(\xi)}{i\hat{\lambda}_{s,\mathrm{ph}}^h(\xi)}\hat{u}^{h,\pm}(\xi)\exp\big(i\xi x_j \pm it\hat{\lambda}_{s,\mathrm{ph}}^h(\xi)\big)\,d\xi, \tag{6.17}$$

with

$$\hat{u}^{h,\pm}(\xi) := \frac{1}{2}\big(i\hat{\lambda}_{s,\mathrm{ph}}^h(\xi)\hat{u}^{h,0}(\xi) \pm \hat{u}^{h,1}(\xi)\big).$$

Then, for all $\xi \in \Pi^h$,

$$|\hat{u}^{h,+}(\xi)|^2 + |\hat{u}^{h,-}(\xi)|^2 = \frac{1}{2}\big(\hat{\Lambda}_{s,\mathrm{ph}}^h(\xi)|\hat{u}^{h,0}(\xi)|^2 + |\hat{u}^{h,1}(\xi)|^2\big). \tag{6.18}$$

As in **Step C2**, we consider the change of variable $\xi \to -\xi$ in the term with the $-$ sign in the right-hand side of (6.17). Then, by using the fact that $\hat{\lambda}_{s,\mathrm{ph}}^h(\xi)$ is an odd function, we have

$$\begin{pmatrix} \{u_s^h\}(x_j,t) \\ [u_s^h](x_j,t) \end{pmatrix} = \frac{1}{2\pi}\sum_{\pm}\int_{\Pi_\delta^h} \frac{\hat{\mathbf{v}}_{s,\mathrm{ph}}^h(\pm\xi)}{i\hat{\lambda}_{s,\mathrm{ph}}^h(\pm\xi)}\hat{u}^{h,\pm}(\pm\xi)\exp\big(it\hat{\lambda}_{s,\mathrm{ph}}^h(\xi) \pm i\xi x_j\big)\,d\xi. \tag{6.19}$$

In order to proceed as in **Step C2** and to identify the Fourier transform in the time variable of the numerical solution in terms of its SDFT in the space variable, we perform the change of variable $\xi \to \hat{\lambda}_{s,\mathrm{ph}}^h(\xi)$. According to property (c4) in Proposition 4.3, $\hat{\lambda}_{s,\mathrm{ph}}^h(\xi)$ is an increasing function in $\xi$, so that it is injective. Consequently, the change of variable $\tau = \hat{\lambda}_{s,\mathrm{ph}}^h(\xi)$ is well defined. Set $\xi(\tau) = (\hat{\lambda}_{s,\mathrm{ph}}^h)^{-1}(\tau)$. Observe that

$$d\xi = \frac{1}{\partial_\xi\hat{\lambda}_{s,\mathrm{ph}}^h(\xi(\tau))}\,d\tau. \tag{6.20}$$

Set

$$\hat{w}_j^{h,+}(\xi) := \sum_{\pm}[\hat{u}^{h,\pm}(\pm\xi)\exp(\pm i\xi x_j)], \quad \hat{w}_j^{h,-}(\xi) := \sum_{\pm}[\pm\hat{u}^{h,\pm}(\pm\xi)\exp(\pm i\xi x_j)],$$

so that, for all $\xi \in \Pi^h$,

$$|\hat{w}_j^{h,+}(\xi)|^2 + |\hat{w}_j^{h,-}(\xi)|^2 = 2\big(|\hat{u}^{h,+}(\xi)|^2 + |\hat{u}^{h,-}(-\xi)|^2\big) \tag{6.21}$$

does not depend anymore on $x_j \in \mathscr{G}^h$, which is the analogue of identity (6.11) in **Step C3**.

Taking into account the fact that both $\hat{\lambda}^h_{s,\mathrm{ph}}(\xi)$ and $\hat{v}^h_{s,\mathrm{ph}}(\xi)$ in (4.8) are odd functions with respect to the wave number $\xi$, after the change of variable $\hat{\lambda}^h_{s,\mathrm{ph}}(\xi) = \tau$, the two components of the solution (6.19) become

$$\{u_s^h\}(x_j,t) = \frac{1}{2\pi}\int_{S_\delta^h} \frac{1}{i\tau}\frac{\hat{w}_j^{h,-}(\xi(\tau))}{\sqrt{1+|\hat{v}^h_{s,\mathrm{ph}}(\xi(\tau))|^2}}\frac{\exp(it\tau)}{\partial_\xi\hat{\lambda}^h_{s,\mathrm{ph}}(\xi(\tau))}\,d\tau \tag{6.22}$$

and

$$[u_s^h](x_j,t) = \frac{1}{2\pi}\int_{S_\delta^h} \frac{1}{i\tau}\frac{\hat{v}^h_{s,\mathrm{ph}}(\xi(\tau))\hat{w}_j^{h,+}(\xi(\tau))}{\sqrt{1+|\hat{v}^h_{s,\mathrm{ph}}(\xi(\tau))|^2}}\frac{\exp(it\tau)}{\partial_\xi\hat{\lambda}^h_{s,\mathrm{ph}}(\xi(\tau))}\,d\tau, \tag{6.23}$$

so that

$$\mathscr{M}^{h,+}\{\partial_t u_s^h\}(x_j,t) = \frac{1}{2\pi}\int_{S_\delta^h} \frac{\cos\left(\frac{\xi(\tau)h}{2}\right)\hat{w}^{h,-}_{j+1/2}(\xi(\tau))}{\sqrt{1+|\hat{v}^h_{s,\mathrm{ph}}(\xi(\tau))|^2}}\frac{\exp(it\tau)}{\partial_\xi\hat{\lambda}^h_{s,\mathrm{ph}}(\xi(\tau))}\,d\tau, \tag{6.24}$$

where $S_\delta^h$ is the set $S_\delta^h := \hat{\lambda}^h_{s,\mathrm{ph}}(\Pi_\delta^h)$ and $\hat{v}^h_{s,ph}(\xi)$ has been introduced in (4.8). Thus, the Fourier transform in time of $\mathscr{M}^{h,\pm}\{\partial_t u_s^h\}(x_j,t)$ is

$$\mathscr{F}_{t\to\tau}(\mathscr{M}^{h,\pm}\{\partial_t u_s^h\})(x_j,\tau) = \frac{\cos\left(\frac{\xi(\tau)h}{2}\right)}{\sqrt{1+|\hat{v}^h_{s,\mathrm{ph}}(\xi(\tau))|^2}}\frac{\hat{w}^{h,-}_{j\pm1/2}(\xi(\tau))}{\partial_\xi\hat{\lambda}^h_{s,\mathrm{ph}}(\xi(\tau))}\chi_{S_\delta^h}(\tau). \tag{6.25}$$

Define

$$d\mu(\xi) := \frac{1}{1+|\hat{v}^h_{s,\mathrm{ph}}(\xi)|^2}\frac{1}{\partial_\xi\hat{\lambda}^h_{s,\mathrm{ph}}(\xi)}\,d\xi.$$

By the Parseval identity (6.16), the definition of $\mathscr{E}^1(\mathbf{u}_s^\mathbf{h}(t))$ in (5.14), and undoing the change of variable $\hat{\lambda}^h_{s,\mathrm{ph}}(\xi) = \tau$, we obtain the explicit expression of $\mathscr{J}^1$ in (6.15) below:

$$\mathscr{J}^1 = \frac{h}{6}\sum_{x_j\in I}\frac{1}{2\pi}\int_{\Pi_\delta^h}\left[|\hat{w}^{h,-}_{j+1/2}(\xi)|^2 + |\hat{w}^{h,-}_{j-1/2}(\xi)|^2\right]\cos^2\left(\frac{\xi h}{2}\right)d\mu(\xi). \tag{6.26}$$

Observe the following identity:

$$\frac{1}{2}\left(|\hat{w}_{j+1/2}^{h,-}(\xi)|^2 + |\hat{w}_{j-1/2}^{h,-}(\xi)|^2\right) = |\hat{w}_j^{h,-}(\xi)|^2 \cos^2\left(\frac{\xi h}{2}\right) + |\hat{w}_j^{h,+}(\xi)|^2 \sin^2\left(\frac{\xi h}{2}\right),$$

so that (6.26) becomes

$$\mathscr{I}^1 = \frac{h}{6\pi} \sum_{x_j \in I} \int_{\Pi_\delta^h} \left[\cos^4\left(\frac{\xi h}{2}\right)|\hat{w}_j^{h,-}(\xi)|^2 + \frac{1}{4}\sin^2(\xi h)|\hat{w}_j^{h,+}(\xi)|^2\right] d\mu(\xi).$$

Using the same arguments as for $\mathscr{I}^1$, we obtain

$$\mathscr{I}^2 = \frac{h}{4\pi} \sum_{x_j \in I} \int_{\Pi_\delta^h} \left[\cos^2\left(\frac{\xi h}{2}\right)|\hat{w}_j^{h,+}(\xi)|^2 + \sin^2\left(\frac{\xi h}{2}\right)|\hat{w}_j^{h,-}(\xi)|^2\right] \frac{\hat{\Lambda}_{1,\mathrm{ph}}^h(\xi)}{\hat{\Lambda}_{s,\mathrm{ph}}^h(\xi)}\, d\mu(\xi),$$

$$\mathscr{I}^3 = \frac{h}{24\pi} \sum_{x_j \in I} \int_{\Pi_\delta^h} \left[\frac{1}{4}\sin^2(\xi h)|\hat{w}_j^{h,-}(\xi)|^2 + \sin^4\left(\frac{\xi h}{2}\right)|\hat{w}_j^{h,+}(\xi)|^2\right] |\hat{v}_{s,\mathrm{ph}}^h(\xi)|^2\, d\mu(\xi),$$

$$\mathscr{I}^4 = \frac{h}{16\pi} \sum_{x_j \in I} \int_{\Pi_\delta^h} \left[\cos^2\left(\frac{\xi h}{2}\right)|\hat{w}_j^{h,-}(\xi)|^2 \right.$$

$$\left. + \sin^2\left(\frac{\xi h}{2}\right)|\hat{w}_j^{h,+}(\xi)|^2\right] \frac{\hat{\Lambda}_{1,\mathrm{ph}}^h(\xi)}{\hat{\Lambda}_{s,\mathrm{ph}}^h(\xi)} |\hat{v}_{s,ph}^h(\xi)|^2\, d\mu(\xi),$$

$$\mathscr{I}^5 = \frac{s-1}{4\pi h} \sum_{x_j \in I} \int_{\Pi_\delta^h} |\hat{w}_j^{h,+}(\xi)|^2 \frac{|\hat{v}_{s,\mathrm{ph}}^h(\xi)|^2}{\hat{\Lambda}_{s,\mathrm{ph}}^h(\xi)}\, d\mu(\xi).$$

and

$$\mathscr{I}^6 + \mathscr{I}^7 = \frac{h}{24\pi} \sum_{x_j \in I} \int_{\Pi_\delta^h} \left(\hat{a}_1^h(\xi)|\hat{w}_j^{h,-}(\xi)|^2 + \hat{a}_2^h(\xi)|\hat{w}_j^{h,+}(\xi)|^2\right) d\mu(\xi),$$

where

$$\hat{a}_1^h(\xi) := \cos^2(\xi h) + \left|1 + \frac{i}{2}\sin(\xi h)\hat{v}_{s,\mathrm{ph}}^h(\xi)\right|^2$$

and

$$\hat{a}_2^h(\xi) := \left|\sin(\xi h) + \frac{i}{2}\hat{v}_{s,\mathrm{ph}}^h(\xi)\right|^2 + \frac{1}{4}|\hat{v}_{s,\mathrm{ph}}^h(\xi)|^2 \cos^2(\xi h).$$

In view of the above expressions for $\mathscr{I}^1, \ldots, \mathscr{I}^5$ and $\mathscr{I}^6 + \mathscr{I}^7$, we conclude that

$$\int_{\mathbb{R}} \mathscr{E}_{s,I}^{W,h}(\mathbf{u}_s^{\mathbf{h}}(t), \partial_t \mathbf{u}_s^{\mathbf{h}}(t))\, dt \le \frac{h}{4\pi} \sum_{x_j \in I} \sum_{\pm} \int_{\Pi_\delta^h} \hat{\alpha}_\pm^h(\xi)|\hat{w}_j^{h,\pm}(\xi)|^2\, d\mu(\xi), \quad (6.27)$$

where

$$\hat{\alpha}_+^h(\xi) = \frac{2}{3}\sin^2\left(\frac{\xi h}{2}\right)\cos^2\left(\frac{\xi h}{2}\right) + \cos^2\left(\frac{\xi h}{2}\right)\frac{\hat{\Lambda}_{1,\mathrm{ph}}^h(\xi)}{\hat{\Lambda}_{s,\mathrm{ph}}^h(\xi)} + \frac{1}{6}|\hat{v}_{s,ph}^h(\xi)|^2\sin^4\left(\frac{\xi h}{2}\right)$$

$$+ \frac{1}{4}|\hat{v}_{s,\mathrm{ph}}^h(\xi)|^2\sin^2\left(\frac{\xi h}{2}\right)\frac{\hat{\Lambda}_{1,\mathrm{ph}}^h(\xi)}{\hat{\Lambda}_{s,\mathrm{ph}}^h(\xi)} + \frac{s-1}{h^2}\frac{|\hat{v}_{s,\mathrm{ph}}^h(\xi)|^2}{\hat{\Lambda}_{s,\mathrm{ph}}^h(\xi)} + \frac{1}{6}\hat{a}_2^h(\xi)$$

and

$$\hat{\alpha}_-^h(\xi) = \frac{2}{3}\cos^4\left(\frac{\xi h}{2}\right) + \sin^2\left(\frac{\xi h}{2}\right)\frac{\hat{\Lambda}_{1,\mathrm{ph}}^h(\xi)}{\hat{\Lambda}_{s,\mathrm{ph}}^h(\xi)} + \frac{1}{6}|\hat{v}_{s,\mathrm{ph}}^h(\xi)|^2\sin^2\left(\frac{\xi h}{2}\right)\cos^2\left(\frac{\xi h}{2}\right)$$

$$+ \frac{1}{4}|\hat{v}_{s,\mathrm{ph}}^h(\xi)|^2\cos^2\left(\frac{\xi h}{2}\right)\frac{\hat{\Lambda}_{1,\mathrm{ph}}^h(\xi)}{\hat{\Lambda}_{s,\mathrm{ph}}^h(\xi)} + \frac{1}{6}\hat{a}_1^h(\xi).$$

In what follows, we will use the following lemma:

**Lemma 6.1.** *For all $\xi \in \Pi^h$, $\hat{\alpha}_+^h(\xi)$, $\hat{\alpha}_-^h(\xi)$ introduced in (6.27) and the symbols $\hat{v}_{s,\mathrm{ph}}^h(\xi)$, $\hat{m}_{s,\mathrm{ph}}^h(\xi)$ introduced in (4.8) and (5.4), the following identity holds:*

$$\hat{\alpha}_+^h(\xi) + \hat{\alpha}_-^h(\xi) = 2\hat{m}_{s,\mathrm{ph}}^h(\xi)\big(1 + |\hat{v}_{s,\mathrm{ph}}^h(\xi)|^2\big).$$

Using the above lemma and (6.21), we obtain

$$|\widehat{w}_j^{h,+}(\xi)|^2\widehat{\alpha}_+^h(\xi) + |\widehat{w}_j^{h,-}(\xi)|^2\widehat{\alpha}_-^h(\xi)$$
$$\le \max\{\widehat{\alpha}_+^h(\xi), \widehat{\alpha}_-^h(\xi)\}\big(|\widehat{w}_j^{h,+}(\xi)|^2 + |\widehat{w}_j^{h,-}(\xi)|^2\big)$$
$$= \left(\frac{\widehat{\alpha}_+^h(\xi) + \widehat{\alpha}_-^h(\xi)}{2} + \frac{|\widehat{\alpha}_+^h(\xi) - \widehat{\alpha}_-^h(\xi)|}{2}\right)\big(|\widehat{w}_j^{h,+}(\xi)|^2 + |\widehat{w}_j^{h,-}(\xi)|^2\big)$$
$$= 2(1 + \widehat{c}_{s,ph}^h(\xi))\widehat{m}_{s,ph}^h(\xi)\big(1 + |\widehat{v}_{s,ph}^h(\xi)|^2\big)\big(|\widehat{u}^{h,+}(\xi)|^2 + |\widehat{u}^{h,-}(-\xi)|^2\big),$$

for all $j \in \mathbb{Z}$, where

$$\hat{c}_{s,\mathrm{ph}}^h(\xi) = \frac{|\hat{\alpha}_+^h(\xi) - \hat{\alpha}_-^h(\xi)|}{2\hat{m}_{s,\mathrm{ph}}^h(\xi)(1 + |\hat{v}_{s,\mathrm{ph}}^h(\xi)|^2)}.$$

Taking into account the fact that the right-hand side of the above inequality does not depend on $j \in \mathbb{Z}$ and that both $\hat{m}_{s,\mathrm{ph}}^h(\xi)$ and $\hat{c}_{s,\mathrm{ph}}^h(\xi)$ are even functions of $\xi$, Lemma 5.1 and (6.18), we obtain

$$\int_{\mathbb{R}} \mathscr{E}_{s,I}^{W,h}(\mathbf{u}_s^{\mathbf{h}}(t), \partial_t \mathbf{u}_s^{\mathbf{h}}(t))\, dt$$
$$\le \frac{1}{2\pi}\int_{\Pi_\delta^h} \frac{1 + \hat{c}_{s,\mathrm{ph}}^h(\xi)}{\partial_\xi \hat{\lambda}_{s,\mathrm{ph}}^h(\xi)}\big(\hat{m}_{s,\mathrm{ph}}^h(\xi)|\hat{u}^{h,1}(\xi) + \hat{r}_{s,\mathrm{ph}}^h(\xi)|\hat{u}^{h,0}(\xi)|^2\big)\, d\xi. \qquad (6.28)$$

By combining (5.4) and (6.28), we obtain the following discrete analogue of (6.7):

$$\int_{\mathbb{R}} \mathscr{E}_{s,I}^{W,h}(\mathbf{u}_s^{\mathbf{h}}(t), \partial_t \mathbf{u}_s^{\mathbf{h}}(t))\, dt \leq 2\mathscr{E}_s^{W,h}(\mathbf{u}^{\mathbf{h},0}, \mathbf{u}^{\mathbf{h},1})(1 + C_{s,\text{ph},\delta})/v^{\star}_{s,\text{ph},\delta},$$

where $v^{\star}_{s,\text{ph},\delta}$ is the minimal group velocity in (6.4) and $C_{s,\text{ph},\delta}$ is the constant in Theorem 6.1 given by

$$C_{s,\text{ph},\delta} := \max_{\xi \in \Pi_\delta^h} \hat{c}^h_{s,\text{ph}}(\xi) = \max_{\eta \in \Pi_\delta^1} \hat{c}^1_{s,\text{ph}}(\eta) \in [0, 1].$$

Observe that both $C_{s,\text{ph},\delta}$ and $v^{\star}_{s,\text{ph},\delta}$, as well as the observability time $T^{\star}_{s,\text{ph},\delta}$ in Theorem 6.1 are independent of $h$. The observability constant $C_s^h(T)$ in (2.9) has the upper bound $C_s^h(T) \leq 1/(T - T^{\star}_{s,\text{ph},\delta})$. This concludes the proof of Theorem 6.1. □

*Remark 6.1.* Some observations on the optimality of the observability time $T^{\star}_{s,\text{ph},\delta}$ in Theorem 6.1 are needed:

- From [9], the expected discrete observability time in Theorem 6.1 is

$$T^{\sharp}_{s,\text{ph},\delta} := \frac{2}{v^{\star}_{s,\text{ph},\delta}}, \tag{6.29}$$

  which is slightly smaller than $T^{\star}_{s,\text{ph},\delta}$ computed by our method. We have, in particular, $T^{\star}_{s,\text{ph},\delta} - T^{\sharp}_{s,\text{ph},\delta} = C_{s,\text{ph},\delta} T^{\sharp}_{s,\text{ph},\delta}$ and $T^{\sharp}_{s,\text{ph},\delta} \leq T^{\star}_{s,\text{ph},\delta} \leq 2T^{\sharp}_{s,\text{ph},\delta}$. However, $T^{\star}_{s,\text{ph},\delta} \to T^{\star} := 2$ as $\delta \to 0$, where $T^{\star}$ is the characteristic time for the continuous model. Also, in view of property (c2) in Proposition 4.3, for all $s \in (1, \infty) \setminus \{3\}$, $T^{\sharp}_{s,\text{ph},\delta}$, $T^{\star}_{s,\text{ph},\delta} \to \infty$ as $\delta \to 1$, which is in accordance with the blowup of the observability constant $C_s^h(T)$ in (2.9) that we discussed in Chap. 5.
- For $s > 3$, one can guarantee the existence of a wave number $\eta_s \in (0, \pi)$ such that the group velocity $\partial_\eta \hat{\lambda}^1_{s,\text{ph}}(\eta)$ is larger than one for $\eta \in (0, \eta_s)$ and smaller than one for $\eta \in [\eta_s, \pi]$. Also, $\eta_s \to \pi$ as $s \to 3$. Then any Fourier truncation of parameter $\delta \in (0, \eta_s/\pi]$ leads to the characteristic velocity $v^{\star}_{s,\text{ph},\delta} = 1$, where $v^{\star}_{s,\text{ph},\delta}$ is the minimal group velocity in (6.4).
- For $s = 3$, the result of Theorem 6.1 holds even for $\delta = 1$ since the physical group velocity $\partial_\xi \hat{\lambda}^h_{3,\text{ph}}(\xi)$ does not vanish at any wave number $\xi \in \Pi^h$.
- The same arguments can be applied to prove Theorem 6.1 for the approximation (2.7) of the Klein–Gordon.

## 6.2 Bi-Grid Filtering on the Physical Mode

In this section, we analyze the observability inequality (2.9) in the class of initial data $(\mathbf{u^{h,0}}, \mathbf{u^{h,1}})$ in (2.6) and (2.7) whose SDFTs take the vectorial form (5.1) and such that for all $i = 0, 1$, $\hat{u}^{h,i}(\xi)$ is the SDFT of a sequence given by a *bi-grid algorithm* of mesh ratio $1/2$. To be more precise, let us define the space of sequences given by this *bi-grid filtering method* to be

$$\mathscr{B}^h := \{\mathbf{f}^h \in \ell^2(\mathscr{G}^h) \text{ such that } f_{2j} = (f_{2j+1} + f_{2j-1})/2, \ \forall j \in \mathbb{Z}\}. \tag{6.30}$$

In other words, any element $\mathbf{f}^h \in \mathscr{B}^h$ is characterized by the fact that only its values on the coarser uniform grid of size $2h$ constituted by the *odd nodal points* $(x_{2j+1})_{j\in\mathbb{Z}}$ are arbitrarily given, whereas the ones at the *even nodes* are obtained by *linear interpolation.*

The main result of this section is the following one:

**Theorem 6.2.** *Consider initial data in (2.6) and (2.7) satisfying (5.1) and such that the scalar function $\hat{u}^{h,i}$ in (5.1) is the SDFT of a sequence $\mathbf{u}^{h,i} \in \mathscr{B}^h$ for all $i = 0, 1$. Then, for all $s > 1$ and all $T > T^\star_{s,\mathrm{ph},1/2}$, with $T^\star_{s,\mathrm{ph},1/2}$ as in Theorem 6.1 for $\delta = 1/2$, the observability inequality (2.9) holds uniformly as $h \to 0$.*

*Proof (of Theorem 6.2).* We proceed as in [37] where, inspired by [41], a *dyadic decomposition argument* has been used in order to prove the uniform boundary observability of the FD semi-discretization of the $2-d$ wave equation in the unit square with initial data given by a bi-grid algorithm. We divide the proof of Theorem 6.2 into several steps as follows:

**Step I. Upper bound of the total energy in terms of the energy concentrated on $\Pi^{2h}$.** For $\Pi^{h,\sharp} \subset \Pi^h$, we define the projection of a sequence $\mathbf{f}^h \in \ell^2(\mathscr{G}^h)$ on $\{\mathbf{g}^h \in \ell^2(\mathscr{G}^h) \text{ s.t. } \mathrm{supp}(\hat{g}^h) \subset \Pi^{h,\sharp}\}$ to be

$$\Gamma_{\Pi^{h,\sharp}} f_j = \frac{1}{2\pi} \int_{\Pi^{h,\sharp}} \hat{f}^h(\xi) \exp(i\xi x_j)\, d\xi. \tag{6.31}$$

The following result (whose proof is postponed to the one of Theorem 6.2) holds:

**Proposition 6.1.** *For all initial data in (2.6) and (2.7) satisfying (5.1) and such that the scalar function $\hat{u}^{h,i}$ in (5.1) is the SDFT of a sequence $\mathbf{u}^{h,i} \in \mathscr{B}^h$, all $i = 0, 1$ and all $s \in (1, \infty)$, there exists a constant $C > 0$, independent of $h$ and $s$, such that*

$$\mathscr{E}_s^{\varsigma,h}(\mathbf{u^{h,0}}, \mathbf{u^{h,1}}) \leq C \mathscr{E}_s^{\varsigma,h}(\Gamma_{\Pi^{2h}} \mathbf{u^{h,0}}, \Gamma_{\Pi^{2h}} \mathbf{u^{h,1}}), \tag{6.32}$$

*where $\varsigma \in \{W, K\}$ and $\Gamma_{\Pi^{2h}}$ is the projection on $\mathscr{I}^h_{1/2}$ in (6.1) with $\delta = 1/2$.*

**Step II. Definition and main properties of the projectors.** For fixed $c > 1$ and $\widehat{\wp} \in C_c^\infty(\mathbb{R})$, we introduce the projector operator $\wp_k : L^2(\mathbb{R}) \to L^2(\mathbb{R})$ as follows:

$$(\wp_k f)(t) = \frac{1}{2\pi} \int_{\mathbb{R}} (\mathscr{F}_{t\to\tau} f)(\tau) \widehat{\wp}\left(\frac{\tau}{c^k}\right) \exp(it\tau)\, d\tau. \tag{6.33}$$

A possible way to construct the projectors $\wp_k$ in (6.33) is to consider four parameters $a, b, c, \mu$ satisfying the restrictions

$$1 < c < \frac{b-\mu}{a+\mu} < \frac{b}{a} < \frac{\hat{\lambda}^h_{s,\mathrm{ph}}((\delta+\epsilon)\pi/h)}{\hat{\lambda}^h_{s,\mathrm{ph}}(\pi\delta/h)}, \tag{6.34}$$

with $\delta = 1/2$, and a function

$$\hat{p} \in C_c^\infty(a,b), \quad 0 \le \hat{p} \le 1, \quad \hat{p} \equiv 1 \text{ in } (a+\mu, b-\mu). \tag{6.35}$$

Then the function $\widehat{\wp}$ generating the projector $\wp_k$ is even, given by

$$\widehat{\wp}(\tau) = \hat{p}(\tau)\chi_{(0,\infty)}(\tau) + \hat{p}(-\tau)\chi_{(-\infty,0)}(\tau). \tag{6.36}$$

In the particular case when the initial data in (2.6) satisfy the condition (5.1), the projectors $\wp_k$ act on the solution $\mathbf{u}^{\mathbf{h}}_s(t)$ as follows:

$$\wp_k \begin{pmatrix} \{u^h_s\}(x_j,t) \\ [u^h_s](x_j,t) \end{pmatrix} = \frac{1}{2\pi}\int_{\Pi^h} \widehat{\wp}\left(\frac{\hat{\lambda}^h_{s,\mathrm{ph}}(\xi)}{c^k}\right) \hat{\mathbf{u}}^h(\xi,t)\exp(i\xi x_j)\,d\xi, \tag{6.37}$$

where $\hat{\mathbf{u}}^h(\xi,t)$ is the particular solution of (2.13) given by (5.3). Thus, for all $k \in \mathbb{N}$, $\wp_k \mathbf{u}^{\mathbf{h}}_s(t)$ is also a solution of (2.6) corresponding to the initial data $(\mathbf{w}^{\mathbf{h},0}, \mathbf{w}^{\mathbf{h},1})$ whose SDFT are

$$\hat{\mathbf{w}}^{h,i}(\xi) = \hat{\mathbf{v}}^h_{s,\mathrm{ph}}(\xi)\widehat{\wp}\left(\frac{\hat{\lambda}^h_{s,\mathrm{ph}}(\xi)}{c^k}\right)\hat{u}^{h,i}(\xi). \tag{6.38}$$

Therefore, the total energy associated with $\wp_k \mathbf{u}^{\mathbf{h}}_s(t)$, $\mathcal{E}^{\varsigma,h}_s(\wp_k \mathbf{u}^{\mathbf{h}}_s(t), \partial_t \wp_k \mathbf{u}^{\mathbf{h}}_s(t))$, is conserved in time for all $\varsigma \in \{W, K\}$. Denote it by $\mathcal{E}^{\varsigma,h}_s(\wp_k \mathbf{u}^{\mathbf{h},0}, \wp_k \mathbf{u}^{\mathbf{h},1})$. On the other hand, the initial data $(\mathbf{w}^{\mathbf{h},0}, \mathbf{w}^{\mathbf{h},1})$ satisfy (5.1), with $\hat{u}^{h,i}(\xi)$ replaced by $\widehat{\wp}(\hat{\lambda}^h_{s,\mathrm{ph}}(\xi)/c^k)\hat{u}^{h,i}(\xi)$. Then, by (5.4), $\mathcal{E}^{\varsigma,h}_s(\wp_k \mathbf{u}^{\mathbf{h},0}, \wp_k \mathbf{u}^{\mathbf{h},1})$ admits the following Fourier representation for $\varsigma = W$ (and a similar expression for $\varsigma = K$):

$$\begin{aligned} &\mathcal{E}^{\varsigma,h}_s(\wp_k \mathbf{u}^{\mathbf{h},0}, \wp_k \mathbf{u}^{\mathbf{h},1}) := \\ &= \frac{1}{4\pi}\int_{\Pi^h} \widehat{\wp}^2\left(\frac{\hat{\lambda}^h_{s,\mathrm{ph}}(\xi)}{c^k}\right)\left(\hat{m}^h_{s,\mathrm{ph}}(\xi)|\hat{u}^{h,1}(\xi)|^2 + \hat{r}^h_{s,\mathrm{ph}}(\xi)|\hat{u}^{h,0}(\xi)|^2\right)d\xi. \end{aligned} \tag{6.39}$$

**Step III. Bounds for the index $k$ of the projectors $\wp_k$.** Since $c > 1$, then $\bigcup_{k=k^\star}^{\infty}(ac^k, bc^k) = (ac^{k^\star}, \infty)$, for all $k^\star \in \mathbb{N}$. Thus, any $\hat{\lambda}^h_{s,\mathrm{ph}}(\xi) > ac^{k^\star}$ is located in the support of at least one projector $\wp_k$, $k \ge k^\star$.

For any $h > 0$ and $\delta = 1/2$, set $k^h$ to be the unique index such that

$$c^{k^h}(a+\mu) \le \hat{\lambda}^h_{s,\mathrm{ph}}(\pi\delta/h) < c^{k^h+1}(a+\mu). \tag{6.40}$$

By using (6.34) and (6.40) with $\delta = 1/2$, for all $k = k^\star, \dots, k^h$, we obtain

$$c^{k^\star}(a+\mu) \le c^k(a+\mu) \le c^{k^h}(a+\mu) \le \hat{\lambda}^h_{s,\mathrm{ph}}(\pi\delta/h) < c^{k^h+1}(a+\mu)$$
$$< c^{k^h}(b-\mu). \tag{6.41}$$

Fix $k^\star \in \mathbb{N}$ independent of $h$ (that will be chosen more precisely later on). Then any frequency $\hat{\lambda}^h_{s,\mathrm{ph}}(\xi) \in \big[(a+\mu)c^{k^\star}, \hat{\lambda}^h_{s,\mathrm{ph}}(\pi\delta/h)\big)$ is contained in at least one interval $((a+\mu)c^k, (b-\mu)c^k)$, $k^\star \le k \le k^h$, i.e., in the region where $\widehat{\wp}(\cdot/c^k) \equiv 1$ and, consequently, for $\delta = 1/2$,

$$1 \le \sum_{k=k^\star}^{k^h} \widehat{\wp}\left(\frac{\hat{\lambda}^h_{s,\mathrm{ph}}(\xi)}{c^k}\right) \qquad \forall \xi \text{ s.t. } \hat{\lambda}^h_{s,\mathrm{ph}}(\xi) \in \big[(a+\mu)c^{k^\star}, \hat{\lambda}^h_{s,\mathrm{ph}}(\pi\delta/h)\big). \tag{6.42}$$

**Step IV. Upper bounds of $\mathscr{E}^{\varsigma,h}_s(\Gamma_{\Pi^{2h}}\mathbf{u}^{\mathbf{h},0}, \Gamma_{\Pi^{2h}}\mathbf{u}^{\mathbf{h},1})$ in terms of the energy of the projectors, $\mathscr{E}^{\varsigma,h}_s(\wp_k\mathbf{u}^{\mathbf{h},0}, \wp_k\mathbf{u}^{\mathbf{h},1})$.** Set

$$\Pi^{h,k^\star} := \{\xi \in \Pi^h : |\hat{\lambda}^h_{s,\mathrm{ph}}(\xi)| \le c^{k^\star}(a+\mu)\}. \tag{6.43}$$

The energy $\mathscr{E}^{\varsigma,h}_s(\Gamma_{\Pi^{2h}}\mathbf{u}^{\mathbf{h},0}, \Gamma_{\Pi^{2h}}\mathbf{u}^{\mathbf{h},1})$ in the right-hand side of (6.32) can be decomposed as follows for all $\varsigma \in \{W, K\}$:

$$\mathscr{E}^{\varsigma,h}_s(\Gamma_{\Pi^{2h}}\mathbf{u}^{\mathbf{h},0}, \Gamma_{\Pi^{2h}}\mathbf{u}^{\mathbf{h},1}) = \mathscr{E}^{\varsigma,h}_s(\Gamma_{\Pi^{h,k^\star}}\mathbf{u}^{\mathbf{h},0}, \Gamma_{\Pi^{h,k^\star}}\mathbf{u}^{\mathbf{h},1})$$
$$+ \mathscr{E}^{\varsigma,h}_s((\Gamma_{\Pi^{2h}} - \Gamma_{\Pi^{h,k^\star}})\mathbf{u}^{\mathbf{h},0}, (\Gamma_{\Pi^{2h}} - \Gamma_{\Pi^{h,k^\star}})\mathbf{u}^{\mathbf{h},1}). \tag{6.44}$$

Using (6.42) and (6.39), we get

$$\mathscr{E}^{\varsigma,h}_s((\Gamma_{\Pi^{2h}} - \Gamma_{\Pi^{h,k^\star}})\mathbf{u}^{\mathbf{h},0}, (\Gamma_{\Pi^{2h}} - \Gamma_{\Pi^{h,k^\star}})\mathbf{u}^{\mathbf{h},1}) \le \sum_{k=k^\star}^{k^h} \mathscr{E}^{\varsigma,h}_s(\wp_k\mathbf{u}^{\mathbf{h},0}, \wp_k\mathbf{u}^{\mathbf{h},1}). \tag{6.45}$$

Therefore, from (6.44) and (6.45), we obtain the following inequality for all $\varsigma \in \{W, K\}$:

$$\mathscr{E}^{\varsigma,h}_s(\Gamma_{\Pi^{2h}}\mathbf{u}^{\mathbf{h},0}, \Gamma_{\Pi^{2h}}\mathbf{u}^{\mathbf{h},1}) \le \mathscr{E}^{\varsigma,h}_s(\Gamma_{\Pi^{h,k^\star}}\mathbf{u}^{\mathbf{h},0}, \Gamma_{\Pi^{h,k^\star}}\mathbf{u}^{\mathbf{h},1}) + \sum_{k=k^\star}^{k^h} \mathscr{E}^{\varsigma,h}_s(\wp_k\mathbf{u}^{\mathbf{h},0}, \wp_k\mathbf{u}^{\mathbf{h},1}). \tag{6.46}$$

**Step V. Bound of the term $\mathscr{E}_s^{K,h}(\Gamma_{\Pi^{h,k^\star}}\mathbf{u}^{\mathbf{h},0}, \Gamma_{\Pi^{h,k^\star}}\mathbf{u}^{\mathbf{h},1})$.** Classical arguments of *semiclassical (Wigner) measures* allow us to get rid of the first term in the right-hand side of (6.46). More precisely, we claim that there exists a constant $C$, uniform as $h \to 0$, such that, for any bi-grid data $(\mathbf{u}^{\mathbf{h},0}, \mathbf{u}^{\mathbf{h},1})$, the following inequality holds:

$$\mathscr{E}_s^{K,h}(\Gamma_{\Pi^{h,k^\star}}\mathbf{u}^{\mathbf{h},0}, \Gamma_{\Pi^{h,k^\star}}\mathbf{u}^{\mathbf{h},1}) \leq C \int_0^T \mathscr{E}_{s,\Omega}^{K,h}(\mathbf{u}^{\mathbf{h},0}, \mathbf{u}^{\mathbf{h},1}, t)\, dt. \tag{6.47}$$

We refer to [30,46–48] for the presentation of the notions of continuous/discrete Wigner transforms and their connections to the so-called Wigner measures through the limit process as the mesh size parameter $h$ goes to zero and also to [13] for the obtention of observability inequalities for the wave equation using pseudo-differential calculus and Wigner measures.

Let us now prove our claim (6.47). We proceed in several steps.

**Step V-a.** First of all, arguing as in the proof of Theorem 6.2 (except for the fifth step), we get

$$\mathscr{E}_s^{K,h}(\mathbf{u}^{\mathbf{h},0}, \mathbf{u}^{\mathbf{h},1}) \leq C_s^1 \int_0^T \mathscr{E}_{s,\Omega}^{K,h}(\mathbf{u}^{\mathbf{h},0}, \mathbf{u}^{\mathbf{h},1}, t)\, dt + C_s^2 \mathscr{E}_s^{K,h}(\Gamma_{\Pi^{h,k^\star}}\mathbf{u}^{\mathbf{h},0}, \Gamma_{\Pi^{h,k^\star}}\mathbf{u}^{\mathbf{h},1}), \tag{6.48}$$

with two constants $C_s^1 = C_s^1(T) > 0$ and $C_s^2 = C_s^2(T) > 0$ independent of $h$. This is a modified observability inequality with a reminder which is precisely the term we would like to estimate in (6.47).

**Step V-b.** To prove (6.47), we argue by contradiction. If (6.47) does not hold, then there exists a sequence of solutions $\mathbf{u}_s^{\mathbf{h}}(t)$ of (2.7) corresponding to bi-grid data such that

$$\mathscr{E}_s^{K,h}(\Gamma_{\Pi^{h,k^\star}}\mathbf{u}^{\mathbf{h},0}, \Gamma_{\Pi^{h,k^\star}}\mathbf{u}^{\mathbf{h},1}) = 1 \text{ and } \int_0^T \mathscr{E}_{s,\Omega}^{K,h}(\mathbf{u}^{\mathbf{h},0}, \mathbf{u}^{\mathbf{h},1}, t)\, dt \to 0 \text{ as } h \to 0. \tag{6.49}$$

**Step V-c.** Combining the previous two steps, we obtain that the total energy $\mathscr{E}_s^{K,h}(\mathbf{u}^{\mathbf{h},0}, \mathbf{u}^{\mathbf{h},1})$ is bounded as $h \to 0$. More precisely, there exists a constant $C_s^3 = C_s^3(T) > 0$ independent of $h$ such that

$$\mathscr{E}_s^{K,h}(\mathbf{u}^{\mathbf{h},0}, \mathbf{u}^{\mathbf{h},1}) \leq C_s^3 \quad \forall h > 0. \tag{6.50}$$

**Step V-d.** From the time conservation of the energy and its Fourier representation formula (2.19), we obtain that the following quantities are bounded in $L^2(\mathbb{R})$ as $h \to 0$ for all $t \geq 0$ and $s > 1$:

$$\hat{u}_t^{h,\{\cdot\}}(\cdot,t),\ \hat{u}_t^{h,[\cdot]}(\cdot,t),\ \sqrt{|\xi|^2+1}\,\hat{u}^{h,\{\cdot\}}(\cdot,t),\ \sqrt{s-\cos^2(\xi h/2)}\,\hat{u}^{h,[\cdot]}(\cdot,t)/h. \tag{6.51}$$

**Step V-e.** From the last bound in (6.51), taking into account that $s > 1$, we get, in particular, uniform bounds in $L^2(\mathbb{R})$ for

$$\frac{\sin(\xi h/2)}{h}\hat{u}^{h,[\cdot]}(\cdot,t) \text{ and } \hat{u}^{h,[\cdot]}(\cdot,t)/h. \tag{6.52}$$

For a sequence $\mathbf{f}^h = (f_j)_{j\in\mathbb{Z}} \in \ell^2(\mathscr{G}^h)$, we define the *piecewise linear and continuous interpolation operator* $(\mathfrak{S}^h\mathbf{f}^h)(x) := \sum_{j\in\mathbb{Z}} f_j\phi_j^{\{\cdot\}}(x)$.

In view of the boundedness properties above and a classical argument relying on the extraction of suitable subsequences, the following limits hold weakly $\star$ as $h \to 0$:

$$\mathfrak{S}^h\mathbf{u}^{h,[\cdot]} \rightharpoonup 0 \text{ in } L^\infty(0,T;H^1(\mathbb{R})) \text{ and } \mathfrak{S}^h\mathbf{u}_t^{h,[\cdot]} \rightharpoonup 0 \text{ in } L^\infty(0,T;L^2(\mathbb{R})).$$

Then classical compactness results ([63, 69]) ensure that, since $H^1(\mathbb{R})$ is compactly embedded in $L^2_{loc}(\mathbb{R})$ (cf. [42], Sect. 8.6, pp. 192), we get the strong convergence

$$\mathfrak{S}^h\mathbf{u}^{h,[\cdot]} \to 0 \text{ in } C^0([0,T];L^2_{loc}(\mathbb{R}))$$

and then $\mathfrak{S}^h\mathbf{u}^{h,[\cdot]}(t) \to 0$ in $L^2_{loc}(\mathbb{R}))$ for all $t \in [0,T]$ and in any $H^r_{loc}(\mathbb{R})$, for any $r \in (0,1)$ and all $t \in [0,T]$. Since $\mathfrak{S}^h\mathbf{u}^{h,[\cdot]}(t)$ is bounded in $H^1(\mathbb{R})$, this implies that $\mathfrak{S}^h\mathbf{u}^{h,[\cdot]}(t) \rightharpoonup 0$ weakly in $H^1(\mathbb{R})$, for all $t \in [0,T]$.

Similarly, using Eq. (2.7) and the Aubin–Lions lemma (cf. [62], Proposition III.1.3, pp. 106), we obtain that $\partial_t(\mathfrak{S}^h\mathbf{u}_t^{h,[\cdot]})$ is bounded in $L^\infty(0,T;H^{-1}(\mathbb{R}))$, so that $\mathfrak{S}^h\mathbf{u}_t^{h,[\cdot]} \to 0$ in $C^0([0,T];H^{-r}_{loc}(\mathbb{R}))$ for all $r > 0$ and $\mathfrak{S}^h\mathbf{u}_t^{h,[\cdot]}(t) \to 0$ in $H^{-r}_{loc}(\mathbb{R})$ and $\mathfrak{S}^h\mathbf{u}_t^{h,[\cdot]}(t) \rightharpoonup 0$ weakly in $L^2(\mathbb{R})$ for all $t \in [0,T]$.

**Step V-f.** Let us use the remaining information in (6.51). After extracting subsequences, we get the boundedness of $\mathfrak{S}^h\mathbf{u}^{h,\{\cdot\}}(t)$ in $H^1(\mathbb{R})$ and of $\mathfrak{S}^h\mathbf{u}_t^{h,\{\cdot\}}(t)$ in $L^2(\mathbb{R})$ for all $t \in [0,T]$, so that

$$\mathfrak{S}^h\mathbf{u}^{h,\{\cdot\}} \rightharpoonup u \text{ weakly} \star \text{ in } L^\infty(0,T;H^1(\mathbb{R}))$$

and

$$\mathfrak{S}^h\mathbf{u}_t^{h,\{\cdot\}} \rightharpoonup u_t \text{ weakly} \star \text{ in } L^\infty([0,T];L^2(\mathbb{R})),$$

where $u(x,t)$ solves the continuous Klein–Gordon equation (1.26) with initial data $(u^0,u^1)$ being the weak limit in $H^1 \times L^2(\mathbb{R})$ of the averages $(\mathbf{u}^{h,\{\cdot\},0},\mathbf{u}^{h,\{\cdot\},1})$ of the initial data in (2.7). Indeed, the convergence in the classical sense of numerical analysis of the SIPG method ensures that $u$ solves the Klein–Gordon equation and vanishes in $\Omega \times (0,T)$. Thus, by the *Holmgren uniqueness theorem* (cf. [44]) and the fact that the time $T$ and the exterior domain $\Omega$ satisfy the *geometric control condition* (GCC), we obtain $u \equiv 0$ in $\mathbb{R} \times [0,T]$. Note that, in the simple $1-d$ setting under consideration, the fact that $u$ is the trivial function can be obtained from the d'Alembert formula.

Arguing similarly to **Step V.e**, we obtain that, for all $r > 0$,

$$\mathfrak{S}^h\mathbf{u}^{h,\{\cdot\}} \to 0 \text{ strongly in } L^\infty(0,T;L^2_{loc}(\mathbb{R}))$$

and

$$\mathfrak{S}^h \mathbf{u}_t^{h,\{\cdot\}} \to 0 \text{ strongly in } L^\infty(0,T;H_{loc}^{-r}(\mathbb{R})).$$

**Step V-g.** Let us remark that since $(\mathfrak{S}^h \mathbf{u}^{h,\{\cdot\}})_x$ is bounded in $L^\infty(0,T;L^2(\mathbb{R}))$ as $h \to 0$, then $|(\mathfrak{S}^h \mathbf{u}^{h,\{\cdot\}})_x|^2$ is bounded in $L^\infty(0,T;L^1(\mathbb{R}))$, so that a subsequence of it converges to a positive Radon measure $\nu$ in $L^\infty(0,T;\mathcal{M}(\mathbb{R}))$ weakly star, where $\mathcal{M}(\mathbb{R})$ is the space of Radon measures on $\mathbb{R}$. It is clear that i) $\nu(\Omega \times (0,T)) = 0$ and ii) $\nu(\mathbb{R} \times (0,T)) = 1$ (due to the first condition in (6.49)). One can show the existence of the so-called Wigner (semiclassical) measure $\mu$ satisfying

$$\int_{[-\pi,\pi]\times\mathbb{R}} \mu(x,t,d\xi,d\tau) = \nu(x,t),$$

such that $\mathrm{supp}(\mu) \subset \{(x,t,\xi,\tau) \in \mathbb{R}^2 \times [-\pi,\pi] \times \mathbb{R}, \tau^2 = \hat{\Lambda}^1_{s,\mathrm{ph}}(\xi)\}$ and $\mu$ propagates along the characteristics $x_{s,\mathrm{ph}}(t) = x^\star \pm t\partial_\xi \hat{\lambda}^1_{s,\mathrm{ph}}(\xi)$ of the DG approximation.

**Step V-h.** Combining the fact that $\mu(\Omega \times (0,T) \times [-\pi/2,\pi/2] \times \mathbb{R}) = 0$, that $T$ and $\Omega$ satisfy the geometric control condition for the discrete dynamics, and that, in the bi-grid class of initial data, the energy of the high-frequency components is bounded in terms of the energy of the low-frequency ones, we obtain that $\mu(\mathbb{R} \times (0,T) \times [-\pi,\pi] \times \mathbb{R}) = 0$, so that $\nu(\mathbb{R} \times (0,T)) = 0$, which is a contradiction with ii).

**Step VI. Both average and jump components of $\wp_k \mathbf{u}^{\mathbf{h},i}$ belong to $\mathcal{I}^h_{1/2+\epsilon}$ for all $i = 0,1$.** From the definition (6.36) of $\hat{\wp}$, we remark that for all $k^\star \le k \le k^h$, the support of the SDFT of $\wp_k \mathbf{u}^{\mathbf{h},i}$, $i = 0,1$, contains only wave numbers $\xi$ such that $\hat{\lambda}^h_{s,\mathrm{ph}}(\xi) \in (ac^k, bc^k)$. Moreover, according to (6.34) and (6.40), the frequencies $\hat{\lambda}^h_{s,\mathrm{ph}}(\xi)$ involved in the projectors $\wp_k \mathbf{u}^{\mathbf{h},i}$, $i = 0,1$, satisfy the inequalities below:

$$\hat{\lambda}^h_{s,\mathrm{ph}}(\xi) \le c^k b < c^{k^h} b \le \hat{\lambda}^h_{s,\mathrm{ph}}(\pi\delta/h)\frac{b}{(a+\mu)} \le \hat{\lambda}^h_{s,\mathrm{ph}}((\delta+\epsilon)\pi/h), \qquad (6.53)$$

with $\delta = 1/2$. In view of the fact that $\hat{\lambda}^h_{s,\mathrm{ph}}(\xi)$ is strictly increasing in $\xi$, we obtain that $|\xi| \le (1/2+\epsilon)\pi/h$. Then $\wp_k \mathbf{u}^{\mathbf{h},i} \in (\mathcal{I}^h_{1/2+\epsilon})^2$ for all $i = 0,1$ and all $k^\star \le k \le k^h$.

**Step VII. Consequences of the continuity of the optimal time $T^\star_{s,\mathrm{ph},\delta}$ in (6.3) with respect to $\delta$.** By the continuous dependence of $T^\star_{s,\mathrm{ph},\delta}$ with respect to $\delta$ and the conservation of the total energy, for all $T > T^\star_{s,\mathrm{ph},\delta}$, there exist $\gamma, \epsilon > 0$ such that $T - 4\gamma > T^\star_{s,\mathrm{ph},\delta+\epsilon} > T^\star_{s,\mathrm{ph},\delta}$ and the following observability inequality holds uniformly as $h \to 0$ for all $\varsigma \in \{W,K\}$:

$$\mathcal{E}^{\varsigma,h}_s(\mathbf{u}^{\mathbf{h},0},\mathbf{u}^{\mathbf{h},1}) \le C_{s,\mathrm{ph},\delta+\epsilon}(T-4\gamma)\int_{2\gamma}^{T-2\gamma} \mathcal{E}^{\varsigma,h}_{s,\Omega}(\mathbf{u}^{\mathbf{h}}_s(t),\partial_t \mathbf{u}^{\mathbf{h}}_s(t))\,dt, \qquad (6.54)$$

for all initial data $(\mathbf{u}^{\mathbf{h},0}, \mathbf{u}^{\mathbf{h},1})$ in (2.7) concentrated on the physical mode, i.e., satisfying (5.1) and such that each scalar function $\hat{u}^{h,i}(\xi)$ in (5.1), $i = 0, 1$, is the SDFT of a sequence $\mathbf{u}^{h,i} \in \mathscr{I}^h_{\delta+\epsilon}$. Since we have proved that $\wp_k \mathbf{u}^{\mathbf{h},i} \in (\mathscr{I}^h_{1/2+\epsilon})^2$, for all $i = 0, 1$, we may apply (6.54) so that, for all $T - 4\gamma > T^\star_{s,\mathrm{ph},1/2+\epsilon}$, the following observability inequality holds for each $k^\star \le k \le k^h$, with $\delta = 1/2$:

$$\mathscr{E}^{\varsigma,h}_s(\wp_k \mathbf{u}^{\mathbf{h},0}, \wp_k \mathbf{u}^{\mathbf{h},1}) \le C_{s,\mathrm{ph},\delta+\epsilon}(T - 4\gamma) \int_{2\gamma}^{T-2\gamma} \mathscr{E}^{\varsigma,h}_{s,\Omega}(\wp_k \mathbf{u}^{\mathbf{h}}_s(t), \partial_t \wp_k \mathbf{u}^{\mathbf{h}}_s(t))\, dt. \tag{6.55}$$

**Step VIII.** In order to conclude the proof of Theorem 6.2, we need to **estimate the energy concentrated in $\Omega$ of the projectors $(\wp_k \mathbf{u}^{\mathbf{h},0}, \wp_k \mathbf{u}^{\mathbf{h},1})$, $k \ge k^\star$, in terms of the energy concentrated in $\Omega$ of the initial data** $(\mathbf{u}^{\mathbf{h},0}, \mathbf{u}^{\mathbf{h},1})$. This can be done by means of the following lemma:

**Lemma 6.2 (cf. [37]).** *Let $\mathscr{H}$ be a Hilbert space, $\|\cdot\|_{\mathscr{H}}$ the associated norm, $\wp_k$ the projectors defined by (6.33), and $\widehat{\wp} \in C^\infty_c(\mathbb{R})$ the function generating the projectors $\wp_k$. For any positive $T$, $\gamma < T/4$ and $c > 1$, there exist positive constants $C(\widehat{\wp}, c)$ and $C^\star(\widehat{\wp}, T, \gamma)$ such that the following inequality holds:*

$$\sum_{k \geqslant k^\star} \int_{2\gamma}^{T-2\gamma} \|\wp_k w(t)\|^2_{\mathscr{H}}\, dt$$
$$\le C(\widehat{\wp}, c) \int_0^T \|w(t)\|^2_{\mathscr{H}}\, dt + \frac{C^\star(\widehat{\wp}, T, \gamma)}{c^{2k^\star}} \sup_{j \in \mathbb{Z}} \|w\|^2_{L^2(lT,(l+1)T,\mathscr{H})}, \tag{6.56}$$

*for all positive integers $k^\star$ and $w \in L^2(\mathbb{R}, \mathscr{H})$.*

For our particular case, we apply Lemma 6.2 with $w(t) = \mathbf{u}^{\mathbf{h}}_s(t)$ and

$$\mathscr{H} := \{(\mathbf{f}^{\mathbf{h},0}, \mathbf{f}^{\mathbf{h},1}) \in (\ell^2(\mathscr{G}^{\mathbf{h}} \cap \Omega))^2 \text{ s.t. } \|(\mathbf{f}^{\mathbf{h},0}, \mathbf{f}^{\mathbf{h},1})\|^2_{\mathscr{H}} := \mathscr{E}^h_{s,\Omega}(\mathbf{f}^{\mathbf{h},0}, \mathbf{f}^{\mathbf{h},1}) < \infty\}.$$

We also consider the following obvious *admissibility inequality* (for all $\varsigma \in \{W, K\}$), which is useful to absorb the last term in the right-hand side of (6.56) into the total energy in the left-hand side of (6.32):

$$\int_{lT}^{(l+1)T} \mathscr{E}^{\varsigma,h}_{s,\Omega}(\mathbf{u}^{\mathbf{h}}_s(t), \partial_t \mathbf{u}^{\mathbf{h}}_s(t))\, dt \le T \mathscr{E}^{\varsigma,h}_s(\mathbf{u}^{\mathbf{h},0}, \mathbf{u}^{\mathbf{h},1}) \quad \forall l \in \mathbb{Z}. \tag{6.57}$$

The lower index of the projectors, $k^\star$, can be chosen large enough but still independent of $h$ such that

$$CTC_{s,\mathrm{ph},1/2+\epsilon}(T - 4\gamma) \frac{C^\star(\widehat{\wp}, T, \gamma)}{c^{2k^\star}} \le \frac{1}{2},$$

where $C$ is the constant introduced in Proposition 6.1. This concludes the proof of Theorem 6.2. □

*Remark 6.2.* Note that the compactness arguments in **Step V** are rigorous for the Klein–Gordon equation. To prove Theorem 6.2 in the case of the wave equation, we can repeat the arguments used for the proof of Theorem 6.1.

The remaining part of this section is devoted to the proof of Proposition 6.1 during which we will apply the following well-known result:

**Lemma 6.3 (cf. [37]).** *The SDFT $\hat{f}^h(\xi)$ of any sequence $\mathbf{f}^h = (f_j)_{j\in\mathbb{Z}}$ belonging to the bi-grid class $\mathscr{B}^h$ introduced in (6.30) satisfies the identity*

$$\hat{f}^h(\xi) = \hat{b}^h(\xi)\hat{f}^{2h}(\xi), \tag{6.58}$$

*where $\hat{b}^h(\xi)$ is the Fourier symbol of the bi-grid algorithm of mesh ratio $1/2$ defined as*

$$\hat{b}^h(\xi) := \cos^2\left(\frac{\xi h}{2}\right) \tag{6.59}$$

*and $\hat{f}^{2h}(\xi)$ is the extension by $\pi/h$-periodicity of the SDFT at scale $2h$ of $\mathbf{f}^{2h} := (f_{2j+1})_{j\in\mathbb{Z}}$.*

*Proof (of Proposition 6.1).* We prove the result for $\varsigma = W$. For $\varsigma = K$, the arguments are similar. The total energy $\mathscr{E}_s^{W,h}(\mathbf{u^{h,0}}, \mathbf{u^{h,1}})$ in (5.4) admits the following decomposition:

$$\mathscr{E}_s^{W,h}(\mathbf{u^{h,0}}, \mathbf{u^{h,1}}) = \mathscr{E}_s^{W,h}(\Gamma_{\Pi^{2h}}\mathbf{u^{h,0}}, \Gamma_{\Pi^{2h}}\mathbf{u^{h,1}}) + \mathscr{E}_s^{W,h}(\Gamma_{\Pi^h\setminus\Pi^{2h}}\mathbf{u^{h,0}}, \Gamma_{\Pi^h\setminus\Pi^{2h}}\mathbf{u^{h,1}}). \tag{6.60}$$

From the fact that the scalar function $\hat{u}^{h,i}$ in (5.1), with $i = 0, 1$, is the SDFT of a sequence $\mathbf{u}^{h,i}$ belonging to the bi-grid class $\mathscr{B}^h$ in (6.30) and from Lemma 6.3, we obtain

$$\hat{u}^{h,i}(\xi) = \hat{b}^h(\xi)\hat{u}^{2h,i}(\xi),\ \xi \in \Pi^h,\ \forall i = 0, 1, \tag{6.61}$$

where both $\hat{u}^{2h,0}(\xi)$ and $\hat{u}^{2h,1}(\xi)$ are $\pi/h$-periodic functions. Therefore,

$$\begin{aligned}&\mathscr{E}_s^{W,h}(\Gamma_{\Pi^h\setminus\Pi^{2h}}\mathbf{u^{h,0}}, \Gamma_{\Pi^h\setminus\Pi^{2h}}\mathbf{u^{h,1}})\\ &= \frac{1}{4\pi}\int_{\Pi^{2h}}\left(\hat{m}_{s,\mathrm{ph}}^h(\xi)\hat{m}_{s,\mathrm{ph}}^{h,\sharp}(\xi)|\hat{u}^{h,1}(\xi)|^2 + \hat{r}_{s,\mathrm{ph}}^h(\xi)\hat{r}_{s,\mathrm{ph}}^{h,\sharp}(\xi)|\hat{u}^{h,0}(\xi)|^2\right)d\xi,\end{aligned}$$

where $\hat{m}_{s,\mathrm{ph}}^h(\xi)$ and $\hat{r}_{s,\mathrm{ph}}^h(\xi)$ are the scalar functions introduced in (5.4) and

$$\hat{m}_{s,\mathrm{ph}}^{h,\sharp}(\xi) := \frac{\hat{m}_{s,\mathrm{ph}}^h(\xi - \mathrm{sign}(\xi)\pi/h)}{\hat{m}_{s,\mathrm{ph}}^h(\xi)}\tan^4\left(\frac{\xi h}{2}\right)$$

and

$$\hat{r}_{s,\mathrm{ph}}^{h,\sharp}(\xi) := \frac{\hat{r}_{s,\mathrm{ph}}^{h}(\xi - \mathrm{sign}(\xi)\pi/h)}{\hat{r}_{s,\mathrm{ph}}^{h}(\xi)} \tan^4\left(\frac{\xi h}{2}\right).$$

In order to conclude the proof of Proposition 6.1, we will show that, for all $s > 1$, the two constants below are uniformly bounded as $h \to 0$, as $s \to \infty$ and as $s \to 1$:

$$c_{s,\mathrm{ph}}^{m,\sharp} := \max_{\xi \in \Pi^{2h}} \hat{m}_{s,\mathrm{ph}}^{h,\sharp}(\xi) \text{ and } c_{s,\mathrm{ph}}^{r,\sharp} := \max_{\xi \in \Pi^{2h}} \hat{r}_{s,\mathrm{ph}}^{h,\sharp}(\xi). \tag{6.62}$$

Thus, the constant $C$ in (6.32) can be taken to be

$$C := 1 + \sup_{s \in (1,\infty)} \max\left\{c_{s,\mathrm{ph}}^{m,\sharp}, c_{s,\mathrm{ph}}^{r,\sharp}\right\}.$$

Let us first observe that by rescaling, the two constants $c_{s,\mathrm{ph}}^{m,\sharp}$ and $c_{s,\mathrm{ph}}^{r,\sharp}$ do not depend on $h$. Thus, in what follows, we restrict ourselves to $h = 1$. Let us show that, for all $s \in (1, \infty)$, we have $\hat{m}_{s,\mathrm{ph}}^{1,\sharp}, \hat{r}_{s,\mathrm{ph}}^{1,\sharp} \in L^\infty(\Pi^2)$.

Both functions $\hat{m}_{s,\mathrm{ph}}^{1}(\xi)$ and $\hat{r}_{s,\mathrm{ph}}^{1}(\xi)$ in (5.4) are positive trigonometric expressions involving the physical eigenvector $\hat{\mathbf{v}}_{s,\mathrm{ph}}^{1}(\xi)$ introduced in (4.7) in the following way:

$$\hat{m}_{s,\mathrm{ph}}^{1}(\xi)(1 + |\hat{v}_{s,\mathrm{ph}}^{1}(\xi)|^2) = \frac{1}{3} + \frac{1}{12}|\hat{v}_{s,\mathrm{ph}}^{1}(\xi)|^2 + \frac{2}{3}\left|\cos\left(\frac{\xi}{2}\right) + \frac{i}{2}\sin\left(\frac{\xi}{2}\right)\hat{v}_{s,\mathrm{ph}}^{1}(\xi)\right|^2 \tag{6.63}$$

and

$$\hat{r}_{s,\mathrm{ph}}^{1}(\xi)(1 + |\hat{v}_{s,\mathrm{ph}}^{1}(\xi)|^2) = 4\sin^2\left(\frac{\xi}{2}\right) + \left(s - \cos^2\left(\frac{\xi}{2}\right)\right)|\hat{v}_{s,\mathrm{ph}}^{1}(\xi)|^2. \tag{6.64}$$

According to Proposition 4.1, the scalar function $\hat{v}_{s,\mathrm{ph}}^{1}(\xi)$ in (4.8) is a continuous trigonometric expression which vanishes at $\xi = 0$ and at $\xi = \pm\pi$ (for $s > 3$) and blows up at $\xi = \pm\pi$ (for $s \in (1, 3)$). Thus, we have to study carefully the behavior of $\hat{m}_{s,\mathrm{ph}}^{1,\sharp}(\xi)$ and $\hat{r}_{s,\mathrm{ph}}^{1,\sharp}(\xi)$ as $\xi \to 0$, since, away from $\xi = 0$, both these functions are bounded. Since

$$\lim_{\xi \to 0} \hat{m}_{s,\mathrm{ph}}^{1}(\xi \pm \pi) = \begin{cases} 1/4, \ s \in (1,3) \\ 2/7, \ s = 3, \\ 1/3, \ s \in (3,\infty), \end{cases}$$

$$\lim_{\xi\to 0} \hat{r}^{1}_{s,\mathrm{ph}}(\xi \pm \pi) = \begin{cases} s, & s \in (1,3) \\ 24/7, & s = 3 \\ 4, & s \in (3,\infty), \end{cases} \tag{6.65}$$

$$\lim_{\xi\to 0} \hat{m}^{1}_{s,\mathrm{ph}}(\xi) = 1, \quad \lim_{\xi\to 0} \hat{r}^{1}_{s,\mathrm{ph}}(\xi) = 0, \text{ but } \lim_{\xi\to 0} \frac{\hat{\Lambda}^{1}_{1,\mathrm{ph}}(\xi)}{\hat{r}^{1}_{s,\mathrm{ph}}(\xi)} = 1, \tag{6.66}$$

we conclude that, for all $s \in (1,\infty)$, we get

$$\lim_{\xi\to 0} \hat{m}^{1,\sharp}_{s,\mathrm{ph}}(\xi) = \lim_{\xi\to 0} \hat{r}^{1,\sharp}_{s,\mathrm{ph}}(\xi) = 0.$$

Consequently, $\hat{m}^{1,\sharp}_{s,\mathrm{ph}}, \hat{r}^{1,\sharp}_{s,\mathrm{ph}} \in L^{\infty}(\Pi^2)$ and the constants $c^{m,\sharp}_{s,\mathrm{ph}}$ and $c^{r,\sharp}_{s,\mathrm{ph}}$ in (6.62) are well defined. Moreover,

$$\lim_{s\to\infty} \hat{m}^{h,\sharp}_{s,\mathrm{ph}}(\xi) = \frac{2-\cos(\xi h)}{2+\cos(\xi h)} \tan^4\left(\frac{\xi h}{2}\right), \quad \lim_{s\to\infty} \hat{r}^{h,\sharp}_{s,\mathrm{ph}}(\xi) = \tan^2\left(\frac{\xi h}{2}\right),$$

$$\lim_{s\to 1} \hat{m}^{h,\sharp}_{s,\mathrm{ph}}(\xi) = \frac{1+3\sin^2\left(\frac{\xi h}{2}\right)}{1+3\cos^2\left(\frac{\xi h}{2}\right)} \tan^4\left(\frac{\xi h}{2}\right)$$

and

$$\lim_{s\to 1} \hat{r}^{h,\sharp}_{s,\mathrm{ph}}(\xi) = \frac{1+3\sin^2\left(\frac{\xi h}{2}\right)}{1+3\cos^2\left(\frac{\xi h}{2}\right)} \tan^2\left(\frac{\xi h}{2}\right).$$

The fact that the four limits above are bounded for $\xi \in \Pi^{2h}$ shows that both constants $c^{m,\sharp}_{s,\mathrm{ph}}$ and $c^{r,\sharp}_{s,\mathrm{ph}}$ in (6.62) are uniformly bounded as $s \to \infty$ or as $s \to 1$. □

## 6.3 Fourier Truncation of the Averages on Data with Null Jumps

In this section and in the following one, the initial data $\mathbf{u}^{\mathbf{h},i} = (\{\mathbf{u}^{h,i}\}, [\mathbf{u}^{h,i}])$ in (2.6) and (2.7), with $i = 0, 1$, have the following common property concerning the jump part $[\mathbf{u}^{h,i}]$:

$$[\mathbf{u}^{h,i}] \equiv 0 \text{ or, equivalently, } \hat{u}^{h,[\cdot],i}(\xi) \equiv 0. \tag{6.67}$$

From a classical numerical analysis point of view, there are several methods to obtain the approximate initial data in the discrete problems (2.6) and (2.7) out of the data of the continuous models (1.18) and (1.26). The most natural one would consist on projecting the continuous data on the discrete space and, thus, in the context of the DG methods, to exploit the full flexibility that the method allows, making the numerical approximations to be discontinuous at the nodes. But this produces initial data with a significant amount of energy concentrated on the spurious numerical high-frequency components. Thus, one needs to handle the approximation of the initial data by suitable filtering techniques.

One of our main contributions of this book is precisely the development and design of efficient filtering mechanisms. Note that, when dealing with observation and control problems, we are concerned not only with the quality of the numerical approximation for fixed initial data but also with the whole class of solutions of the generated discrete dynamics and, in particular, the solutions generated by very oscillatory data, giving rise to spurious solutions propagating at very low speeds.

Compared to the classical $P_1$-FEM, a DG solution could have two sources of oscillations: one generated by the average components and one by the jump components. The filtering strategies we design are precisely oriented to attenuate these two sources of spurious oscillations. The oscillations on averages are of the same nature as those encountered for the classical $P_1$-FEM scheme and can be efficiently resolved by the classical Fourier truncation or bi-grid techniques ([27, 28, 70]). Oscillations on jumps are intrinsic to DG methods and a very natural strategy to attenuate them consists in completely eliminating them in the initial data. This motivates the choice of initial data we do in the present and the next sections.

The total energy corresponding to this choice of the initial data gets simplified with respect to (2.15), so that it coincides with the one of the $P_1$-FEM approximation (1.13) with initial data $(\mathbf{u}^{h,0}, \mathbf{u}^{h,1})$ of SDFT $(\hat{u}^{h,\{\cdot\},0}, \hat{u}^{h,\{\cdot\},1})$. More precisely,

$$\begin{aligned}\mathscr{E}_s^{W,h}(\mathbf{u}^{\mathbf{h},0}, \mathbf{u}^{\mathbf{h},1}) \\ = \frac{1}{4\pi}\int_{\Pi^h}\left[\frac{2+\cos(\xi h)}{3}|\hat{u}^{h,\{\cdot\},1}(\xi)|^2 + \hat{\Lambda}^h_{1,\mathrm{ph}}(\xi)|\hat{u}^{h,\{\cdot\},0}(\xi)|^2\right]d\xi, \quad (6.68)\end{aligned}$$

with $\hat{\Lambda}^h_{1,\mathrm{ph}}(\xi)$ being the Fourier symbol of the FD approximation in (1.17). A similar Fourier representation holds for $\mathscr{E}_s^{K,h}(\mathbf{u}^{\mathbf{h},0}, \mathbf{u}^{\mathbf{h},1})$.

Then the expression (4.12) of the solution of (2.13) gets simplified as follows:

$$\hat{\mathbf{u}}_s^h(\xi,t) = \hat{\mathbf{V}}_s^{h,\sharp}(\xi)\begin{pmatrix}\hat{u}^h_{s,\mathrm{ph}}(\xi,t)\\ \hat{u}^h_{s,\mathrm{sp}}(\xi,t)\end{pmatrix}, \quad (6.69)$$

where the matrix $\hat{\mathbf{V}}_s^{h,\sharp}(\xi)$ below depends on the components $\hat{v}^h_{s,\mathrm{ph}}(\xi)$ and $\hat{v}^h_{s,\mathrm{sp}}(\xi)$ of the physical and spurious eigenvectors introduced in (4.8):

$$\hat{\mathbf{V}}_s^{h,\sharp}(\xi) := \frac{1}{1-\hat{v}^h_{s,\mathrm{ph}}(\xi)\hat{v}^h_{s,\mathrm{sp}}(\xi)}\begin{pmatrix}1 & -\hat{v}^h_{s,\mathrm{ph}}(\xi)\hat{v}^h_{s,\mathrm{sp}}(\xi)\\ \hat{v}^h_{s,\mathrm{ph}}(\xi) & -\hat{v}^h_{s,\mathrm{ph}}(\xi)\end{pmatrix}$$

and, for $\alpha \in \{\text{ph}, \text{sp}\}$,

$$\hat{u}^h_{s,\alpha}(\xi, t) := \frac{1}{2} \sum_{\pm} \left[ \hat{u}^{h,\{\cdot\},0}(\xi) \pm \frac{\hat{u}^{h,\{\cdot\},1}(\xi)}{i\hat{\lambda}^h_{s,\alpha}(\xi)} \right] \exp(\pm i t \hat{\lambda}^h_{s,\alpha}(\xi)).$$

The main result of this section reads as follows:

**Theorem 6.3.** *Consider initial data* $\mathbf{u}^{\mathbf{h},i} = (\{\mathbf{u}^{h,i}\}, [\mathbf{u}^{h,i}])$ *in (2.6) and (2.7) having null jump components, i.e., satisfying the requirement (6.67) for all* $i = 0, 1$, *and such that the average component* $\{\mathbf{u}^{h,i}\}$ *belongs to the truncated space* $\mathcal{I}^h_\delta$ *in (6.1) with* $\delta \in (0, 1)$. *Then, for all* $T > T^\star_{s,\text{ph},\delta}$, *with* $T^\star_{s,\text{ph},\delta}$ *as in Theorem 6.1, and for all* $s > 3/2$, *the observability inequality (2.9) holds uniformly as* $h \to 0$.

*Remark 6.3.* The proof of Theorem 6.3 based on a dyadic decomposition argument is not valid for all $s > 1$, but for the more restrictive range $s > 3/2$. This constraint on $s$ is motivated by the fact that $\hat{\lambda}^h_{s,\text{sp}}(\xi)$ must not belong to the support of any projectors $\wp_k$ defined by (6.33) for all $k^\star \leq k \leq k^h$, with $k^h$ given by (6.40). A more precise definition of the penalty parameter $s$ for which Theorem 6.3 is valid is as follows:

$$\min_{\xi \in \Pi^h_\delta} |\hat{\lambda}^h_{s,\text{sp}}(\xi)| > \max_{\xi \in \Pi^h_\delta} |\hat{\lambda}^h_{s,\text{ph}}(\xi)|. \tag{6.70}$$

In what follows, we show that for $s > 3/2$ the spurious branch is at the top of the whole physical diagram, i.e., (6.70) holds with $\delta = 1$.

For $s \geq 2$, the spurious diagram $\hat{\lambda}^h_{s,\text{sp}}(\xi)$ has its minimum point located at $\xi = \pi/h$, so that

$$\min_{\xi \in \Pi^h} |\hat{\lambda}^h_{s,\text{sp}}(\xi)| = |\hat{\lambda}^h_{s,\text{sp}}(\pi/h)| = \frac{2}{h}\sqrt{\max\{s, 3\}}$$
$$\geq \max_{\xi \in \Pi^h} |\hat{\lambda}^h_{s,\text{ph}}(\xi)| = |\hat{\lambda}^h_{s,\text{ph}}(\pi/h)| = \frac{2}{h}\sqrt{\min\{s, 3\}}. \tag{6.71}$$

For $s \in (1, 2)$, the spurious diagram $\hat{\lambda}^h_{s,\text{sp}}(\xi)$ has its minimum point located at $\xi = 0$, so that (6.70) requires to find those values of $s$ such that

$$\min_{\xi \in \Pi^h} |\hat{\lambda}^h_{s,\text{sp}}(\xi)| = |\hat{\lambda}^h_{s,\text{sp}}(0)| = \frac{2}{h}\sqrt{3(s-1)}$$
$$> \max_{\xi \in \Pi^h} |\hat{\lambda}^h_{s,\text{ph}}(\xi)| = |\hat{\lambda}^h_{s,\text{ph}}(\pi/h)| = \frac{2}{h}\sqrt{s}, \tag{6.72}$$

which leads precisely to $s > 3/2$.

Before giving the proof of Theorem 6.3, let us introduce some useful notions.

Let $\mathbf{f}^h(t)$ be a time-dependent vector whose SDFT,

$$\hat{f}^h(\xi,t) = \hat{f}^h_{\mathrm{ph}}(\xi)\exp(it\hat{\lambda}^h_{s,\mathrm{ph}}(\xi)) + \hat{f}^h_{\mathrm{sp}}(\xi)\exp(it\hat{\lambda}^h_{s,\mathrm{sp}}(\xi)),\ \xi \in \Pi^h, \tag{6.73}$$

involves both physical and spurious branches of the dispersion diagram and $\hat{f}^h_{\mathrm{ph}}(\xi)$ and $\hat{f}^h_{\mathrm{sp}}(\xi)$ are scalar functions. We define the projection of $\mathbf{f}^h(t)$ on the physical branch as follows:

$$\Gamma_{\mathrm{ph}} f_j(t) = \frac{1}{2\pi}\int_{\Pi^h} \hat{f}^h_{\mathrm{ph}}(\xi)\exp(it\hat{\lambda}^h_{s,\mathrm{ph}}(\xi))\exp(i\xi x_j)\,d\xi. \tag{6.74}$$

The projection $\Gamma_{\mathrm{ph}}$ acts on a sequence $\mathbf{f^h}(t) = (\{\mathbf{f}^h(t)\}, [\mathbf{f}^h(t)])$ associating two values to any grid point $x_j \in \mathscr{G}^h$ as $\Gamma_{\mathrm{ph}}\mathbf{f^h}(t) := (\Gamma_{\mathrm{ph}}\{\mathbf{f}^h(t)\}, \Gamma_{\mathrm{ph}}[\mathbf{f}^h(t)])$.

Particularizing the definition (6.74) on the solution $\mathbf{u}^{\mathbf{h}}_s(t)$ whose SDFT is given by (6.69), we obtain

$$\Gamma_{\mathrm{ph}}\begin{pmatrix}\{u^h_s\}(x_j,t)\\ [u^h_s](x_j,t)\end{pmatrix} = \frac{1}{2\pi}\int_{\Pi^h}\frac{\hat{\mathbf{v}}^h_{s,\mathrm{ph}}(\xi)\sqrt{1+|\hat{v}^h_{s,\mathrm{ph}}(\xi)|^2}}{1-\hat{v}^h_{s,\mathrm{ph}}(\xi)\hat{v}^h_{s,\mathrm{sp}}(\xi)}\hat{u}^h_{s,\mathrm{ph}}(\xi,t)\exp(i\xi x_j)\,d\xi, \tag{6.75}$$

where $\hat{\mathbf{v}}^h_{s,\mathrm{ph}}(\xi)$ is the physical eigenvector introduced in (4.7). Note that $\Gamma_{\mathrm{ph}}\mathbf{u}^{\mathbf{h}}_s(t)$ in (6.75) verifies the SIPG approximation of the wave Eq. (2.6) with initial data $(\mathbf{w}^{\mathbf{h},0}, \mathbf{w}^{\mathbf{h},1})$ whose SDFT is

$$\hat{\mathbf{w}}^{h,i}(\xi) = \hat{\mathbf{v}}^h_{s,\mathrm{ph}}(\xi)\frac{\sqrt{1+|\hat{v}^h_{s,\mathrm{ph}}(\xi)|^2}}{1-\hat{v}^h_{s,\mathrm{ph}}(\xi)\hat{v}^h_{s,\mathrm{sp}}(\xi)}\hat{u}^{h,\{\cdot\},i}(\xi),\ i=0,1. \tag{6.76}$$

Moreover, both $\mathbf{w}^{\mathbf{h},i}$, $i = 0,1$, are concentrated only on the physical mode since they satisfy (5.1) with

$$\hat{u}^{h,i}(\xi) := \frac{\sqrt{1+|\hat{v}^h_{s,\mathrm{ph}}(\xi)|^2}}{1-\hat{v}^h_{s,\mathrm{ph}}(\xi)\hat{v}^h_{s,\mathrm{sp}}(\xi)}\hat{u}^{h,\{\cdot\},i}(\xi),\ i=0,1.$$

Thus, the total energy of $\Gamma_{\mathrm{ph}}\mathbf{u}^{\mathbf{h}}_s(t)$ that we denote by $\mathscr{E}^{W,h}_s(\Gamma_{\mathrm{ph}}\mathbf{u}^{\mathbf{h},0}, \Gamma_{\mathrm{ph}}\mathbf{u}^{\mathbf{h},1})$ is conserved in time and, taking (5.4) into account, it is given by

$$\mathscr{E}^{W,h}_s(\Gamma_{\mathrm{ph}}\mathbf{u}^{\mathbf{h},0}, \Gamma_{\mathrm{ph}}\mathbf{u}^{\mathbf{h},1}) = \frac{1}{4\pi}\int_{\Pi^h}\frac{1+|\hat{v}^h_{s,\mathrm{ph}}(\xi)|^2}{|1-\hat{v}^h_{s,\mathrm{ph}}(\xi)\hat{v}^h_{s,\mathrm{sp}}(\xi)|^2}\Big(\hat{m}^h_{s,\mathrm{ph}}(\xi)|\hat{u}^{h,\{\cdot\},1}(\xi)|^2 + \hat{r}^h_{s,\mathrm{ph}}(\xi)|\hat{u}^{h,\{\cdot\},0}(\xi)|^2\Big)\,d\xi. \tag{6.77}$$

*Proof (of Theorem 6.3).* The proof of the uniform observability in the present case, in which data are preconditioned so that their jumps vanish, follows the same perturbation methods of Sect. 6.2. The null jump condition (6.67) nearly concentrates the whole energy of the data on the physical component of the Fourier decomposition. Thus, it is very likely that the observability results on the solutions concentrated on the physical branch will also hold for the class of solutions under consideration, i.e., with null jumps on the initial datum. Of course, because of the lack of observability of the high-frequency components of the physical branch, one still needs an added filtering mechanism. In this section, we impose the null jump condition with Fourier truncation. In the following one, the Fourier truncation will be replaced by a bi-grid preprocessing.

The proof requires a dyadic decomposition argument and we will emphasize here only the differences with respect to the similar argument in Sect. 6.2. Thus, Proposition 6.1 in **Step I** is replaced by the following one, whose proof will be postponed to the proof of Theorem 6.3:

**Proposition 6.2.** *Consider initial data* $\mathbf{u}^{\mathbf{h},i} = (\{\mathbf{u}^{h,i}\}, [\mathbf{u}^{h,i}])$, $i = 0, 1$, *in (2.6) and (2.7) having null jump components, i.e., satisfying (6.67), and such that the average components* $\{\mathbf{u}^{h,i}\}$ *belong to* $\mathcal{I}_\delta^h$ *introduced in (6.1), for some* $\delta \in (0,1)$. *Then, for all* $s > 1$, *there exists a constant* $C(\delta) > 0$ *uniformly bounded as* $h \to 0$ *and as* $s \to \infty$ *such that the following inequality holds:*

$$\mathscr{E}_s^{\varsigma,h}(\mathbf{u}^{\mathbf{h},0}, \mathbf{u}^{\mathbf{h},1}) \leq C(\delta)\mathscr{E}_s^{\varsigma,h}(\Gamma_{\mathrm{ph}}\mathbf{u}^{\mathbf{h},0}, \Gamma_{\mathrm{ph}}\mathbf{u}^{\mathbf{h},1}) \quad \forall \varsigma \in \{W, K\}. \tag{6.78}$$

The projectors $\wp_k$ in **Step II** are constructed in a similar way. The only difference is that the parameters $a, b, c$ and $\mu$ satisfy (6.34) in which the last inequality is replaced by (6.79) below. In the previous section, that last inequality was needed in **Step VI** to show that the projectors of the solution, $\wp_k \mathbf{u}_s^{\mathbf{h}}(t)$, belong to $(\mathcal{I}_{\delta+\epsilon}^h)^2$. But now $\wp_k \mathbf{u}_s^{\mathbf{h}}(t) \in (\mathcal{I}_{\delta+\epsilon}^h)^2$ for free since the solution itself $\mathbf{u}_s^{\mathbf{h}}(t)$ and its projectors $\wp_k \mathbf{u}_s^{\mathbf{h}}(t)$ belong to $(\mathcal{I}_\delta^h)^2$, which is a subset of $(\mathcal{I}_{\delta+\epsilon}^h)^2$ for any $\epsilon > 0$.

Another technical assumption is the fact that no spurious frequency $\hat{\lambda}_{s,\mathrm{sp}}^h(\xi)$ should belong to the support of the projectors $\wp_k \mathbf{u}_s^{\mathbf{h}}(t)$ for $k \leq k^h$, where $k^h$ is the one given by (6.40). This is equivalent to requiring

$$bc^{k^h} < \min_{\xi \in \Pi_\delta^h} \hat{\lambda}_{s,\mathrm{sp}}^h(\xi). \tag{6.79}$$

If (6.79) holds, then the projectors $\wp_k$ for $k^\star \leq k \leq k^h$ act on the solution $\mathbf{u}_s^{\mathbf{h}}(t)$ in (6.69) as follows:

$$\wp_k \mathbf{u}_s^{\mathbf{h}}(t) = \frac{1}{2\pi}\int_{\Pi_\delta^h} \hat{\mathbf{v}}_{s,\mathrm{ph}}^h(\xi) \frac{\sqrt{1 + |\hat{v}_{s,\mathrm{ph}}^h(\xi)|^2}}{1 - \hat{v}_{s,\mathrm{ph}}^h(\xi)\hat{v}_{s,\mathrm{sp}}^h(\xi)} \widehat{\wp}\left(\frac{\hat{\lambda}_{s,\mathrm{ph}}^h(\xi)}{c^k}\right) \hat{u}_{s,\mathrm{ph}}^h(\xi, t)\exp(i\xi x_j)\, d\xi, \tag{6.80}$$

where $\hat{\mathbf{v}}^h_{s,\mathrm{ph}}(\xi)$ is the physical eigenvector in (4.7). Due to the fact that (6.80) is a solution of (2.6) concentrated on the physical mode (i.e., the corresponding initial data satisfy (5.1)), the total energy of the projectors is conserved in time and, following (5.4), it is given by

$$\mathscr{E}^{W,h}_s(\wp_k \mathbf{u}^{\mathbf{h},0}, \wp_k \mathbf{u}^{\mathbf{h},1}) = \frac{1}{4\pi} \int_{\Pi^h_\delta} \frac{1 + |\hat{v}^h_{s,\mathrm{ph}}(\xi)|^2}{|1 - \hat{v}^h_{s,\mathrm{ph}}(\xi)\hat{v}^h_{s,\mathrm{sp}}(\xi)|^2} \widehat{\wp}^2 \left( \frac{\hat{\lambda}^h_{s,\mathrm{ph}}(\xi)}{c^k} \right)$$
$$\times \left( \hat{m}^h_{s,\mathrm{ph}}(\xi) |\hat{u}^{h,\{\cdot\},1}(\xi)|^2 + \hat{r}^h_{s,\mathrm{ph}}(\xi) |\hat{u}^{h,\{\cdot\},0}(\xi)|^2 \right) d\xi. \tag{6.81}$$

A similar expression can be obtained for $\mathscr{E}^{K,h}_s(\wp_k \mathbf{u}^{\mathbf{h},0}, \wp_k \mathbf{u}^{\mathbf{h},1})$.

Observe that the total energy of the projection on the physical mode of $\mathbf{u}^{\mathbf{h}}_s(t)$ in (6.69) can be split as follows for all $\varsigma \in \{W, K\}$:

$$\mathscr{E}^{\varsigma,h}_s(\Gamma_{\mathrm{ph}} \mathbf{u}^{\mathbf{h},0}, \Gamma_{\mathrm{ph}} \mathbf{u}^{\mathbf{h},1}) = \mathscr{E}^{\varsigma,h}_s(\Gamma_{\Pi^{h,k^\star}} \Gamma_{\mathrm{ph}} \mathbf{u}^{\mathbf{h},0}, \Gamma_{\Pi^{h,k^\star}} \Gamma_{\mathrm{ph}} \mathbf{u}^{\mathbf{h},1})$$
$$+ \mathscr{E}^{\varsigma,h}_s((\Gamma_{\Pi^h_\delta} - \Gamma_{\Pi^{h,k^\star}}) \Gamma_{\mathrm{ph}} \mathbf{u}^{\mathbf{h},0}, (\Gamma_{\Pi^h_\delta} - \Gamma_{\Pi^{h,k^\star}}) \Gamma_{\mathrm{ph}} \mathbf{u}^{\mathbf{h},1}), \tag{6.82}$$

where $\Pi^{h,k^\star}$ is as in (6.43). Identity (6.82) is the analogue of (6.44) in **Step IV**. The first term in the right-hand side can be absorbed by the compactness arguments in **Step V**. Taking into account that the energies (6.77) and (6.81) have the same nature, we can apply (6.42) to obtain the following analogue of (6.45) for all $\varsigma \in \{W, K\}$:

$$\mathscr{E}^{\varsigma,h}_s((\Gamma_{\Pi^h_\delta} - \Gamma_{\Pi^{h,k^\star}}) \Gamma_{\mathrm{ph}} \mathbf{u}^{\mathbf{h},0}, (\Gamma_{\Pi^h_\delta} - \Gamma_{\Pi^{h,k^\star}}) \Gamma_{\mathrm{ph}} \mathbf{u}^{\mathbf{h},1}) \leq \sum_{k=k^\star}^{k^h} \mathscr{E}^{\varsigma,h}_s(\wp_k \mathbf{u}^{\mathbf{h},0}, \wp_k \mathbf{u}^{\mathbf{h},1}). \tag{6.83}$$

To conclude the proof, we highlight the relation between (6.70) and (6.79). From (6.40), we see that a sufficient condition to guarantee (6.79) is requiring the following upper bound of $b$ to hold:

$$b < (a + \mu) \min_{\xi \in \Pi^h_\delta} \hat{\lambda}^h_{s,\mathrm{sp}}(\xi) / \max_{\xi \in \Pi^h_\delta} \hat{\lambda}^h_{s,\mathrm{ph}}(\xi). \tag{6.84}$$

Due to (6.70), we see that it is possible to choose $b$ such that both $a + \mu < b$ and (6.84) hold. The rest of the proof follows the one of Theorem 6.2. □

*Proof (of Proposition 6.2).* Comparing the two total energies involved in (6.78), (6.68) and (6.77) (in both of them, $\Pi^h$ has to be replaced with $\Pi^h_\delta$ since $\{\mathbf{u}^{h,i}\} \in \mathscr{I}^h_\delta$), we see that a sufficient condition to guarantee (6.78) is that the following two functions

$$\hat{m}^{h,\natural}_s(\xi) := \frac{(2 + \cos(\xi h))|1 - \hat{v}^h_{s,\mathrm{ph}}(\xi)\hat{v}^h_{s,\mathrm{sp}}(\xi)|^2}{3\hat{m}^h_{s,\mathrm{ph}}(\xi)(1 + |\hat{v}^h_{s,\mathrm{ph}}(\xi)|^2)}$$

and

$$\hat{r}_s^{h,\natural}(\xi) := \frac{\hat{\Lambda}_{1,\mathrm{ph}}^h(\xi)|1 - \hat{v}_{s,\mathrm{ph}}^h(\xi)\hat{v}_{s,\mathrm{sp}}^h(\xi)|^2}{\hat{r}_{s,\mathrm{ph}}^h(\xi)(1 + |\hat{v}_{s,\mathrm{ph}}^h(\xi)|^2)} \tag{6.85}$$

are uniformly bounded as $h \to 0$ and as $s \to \infty$ on $\xi \in \Pi_\delta^h$. Set

$$c_{s,\delta}^{m,\natural} := \max_{\xi \in \Pi_\delta^h} \hat{m}_s^{h,\natural}(\xi) \text{ and } c_{s,\delta}^{r,\natural} := \max_{\xi \in \Pi_\delta^h} \hat{r}_s^{h,\natural}(\xi). \tag{6.86}$$

By scaling arguments, we see that the two constants in (6.86) are independent of $h$, so that we may restrict ourselves to the case $h = 1$. Then the constant $C(\delta)$ in (6.78) can be taken as

$$C(\delta) := \sup_{s \in (1,\infty)} \max\{c_{s,\delta}^{m,\natural}, c_{s,\delta}^{r,\natural}\}.$$

Observe that $\hat{m}_s^{1,\natural}(\xi)$ and $\hat{r}_s^{1,\natural}(\xi)$ are positive rational expressions depending on trigonometric functions and on $s$. So they are continuous in $s$ and $\xi$, except eventually for the points where the different factors vanish or blow up. These singular points could be located only at $\xi = 0$ or at $\xi = \pi$. Let us observe that taking the limits (6.65) and (6.66) into account, both factors $(2 + \cos(\xi))/3\hat{m}_{s,\mathrm{ph}}^1(\xi)$ and $\hat{\Lambda}_{s,\mathrm{ph}}^1(\xi)/\hat{r}_{s,\mathrm{ph}}^1(\xi)$ are uniformly bounded for $\xi \in \Pi^1$, as $s \to \infty$ or as $s \to 1$. It remains to study the behavior of the expression

$$\hat{v}_s^{1,\natural}(\xi) := \frac{|1 - \hat{v}_{s,\mathrm{ph}}^1(\xi)\hat{v}_{s,\mathrm{sp}}^1(\xi)|^2}{1 + |\hat{v}_{s,\mathrm{ph}}^1(\xi)|^2}. \tag{6.87}$$

Using the expressions of $\hat{v}_{s,\mathrm{ph}}^1(\xi)$ and $\hat{v}_{s,\mathrm{sp}}^1(\xi)$ in (4.8), we see that $-\hat{v}_{s,\mathrm{ph}}^1(\xi)\hat{v}_{s,\mathrm{sp}}^1(\xi) \geq 0$, for all $\xi \in \Pi^1$. Furthermore,

$$\lim_{\xi \to 0} \hat{v}_{s,\mathrm{ph}}^1(\xi)\hat{v}_{s,\mathrm{sp}}^1(\xi) = 0 \quad \forall s \geq 1,$$

and

$$-\lim_{\xi \to \pi} \hat{v}_{s,\mathrm{ph}}^1(\xi)\hat{v}_{s,\mathrm{sp}}^1(\xi) = \begin{cases} +\infty, & s \in [1,3) \\ 1, & s = 3 \\ 0, & s > 3. \end{cases} \tag{6.88}$$

Moreover, for $s \in [1,3)$,

$$-\lim_{\xi \to \pi} \hat{v}_{s,\mathrm{ph}}^1(\xi)\hat{v}_{s,\mathrm{sp}}^1(\xi)\cos^2\left(\frac{\xi}{2}\right) = \frac{(3-s)^2}{4s} \text{ and } \lim_{\xi \to \pi} |\hat{v}_{s,\mathrm{ph}}^1(\xi)|^2\cos^2\left(\frac{\xi}{2}\right) = \frac{(3-s)^2}{s^2}.$$

Consequently, for all $s \in [1,3)$, $\hat{v}_s^{1,\natural}(\xi)$ in (6.87) blows up at $\xi = \pi$:

$$\lim_{\xi \to \pi} \hat{v}_s^{1,\natural}(\xi) \cos^2\left(\frac{\xi}{2}\right) = \frac{(3-s)^2}{16}.$$

However, when $\xi \in \Pi_\delta^1$, with $\delta \in (0,1)$, the function $\hat{v}_s^{1,\natural}(\xi)$ does not blow up.

From the fact that $\lim_{s\to\infty} \hat{v}_s^{1,\natural}(\xi) = 1$ for all $\xi \in \Pi^1$, we deduce that the constants $c_{s,\delta}^{m,\natural}$ and $c_{s,\delta}^{r,\natural}$ in (6.86) are uniformly bounded as $s \to \infty$.

In conclusion, for all $s \in [1,3)$, the constants $c_{s,\delta}^{m,\natural}$ and $c_{s,\delta}^{r,\natural}$ in (6.86) are well defined for all $\delta \in (0,1)$ and blow up as $\delta \to 1$, while for $s \geq 3$ they are well defined even for $\delta = 1$. This does not mean that the Fourier truncation is not necessary for $s \geq 3$, but that it is not required for the proof of Proposition 6.2. However, it is essential when applying **Step VII** in the dyadic decomposition argument. This concludes the proof of Proposition 6.2. □

## 6.4 Bi-Grid Filtering of Averages on Data with Null Jumps

In this section, we deal with initial data $\mathbf{u^{h,\mathit{i}}} = (\{\mathbf{u}^{h,i}\}, [\mathbf{u}^{h,i}])$ in (2.6) or (2.7), with $i = 0, 1$, which, apart from the null jump condition (6.67), satisfy the fact that the average part $\{\mathbf{u}^{h,i}\}$ is obtained by a *bi-grid algorithm* of mesh ratio $1/2$. The main result of this section is as follows (see also the preliminary paper [50]):

**Theorem 6.4.** *Consider initial data* $\mathbf{u^{h,\mathit{i}}} = (\{\mathbf{u}^{h,i}\}, [\mathbf{u}^{h,i}])$ *in (2.6) and (2.7), with* $i = 0, 1$, *satisfying the null jump condition (6.67) and such that each* $\{\mathbf{u}^{h,i}\}$ *belongs to the bi-grid class* $\mathscr{B}^h$ *in (6.30). Then, for all* $T > T^\star_{s,\mathrm{ph},1/2}$, *with* $T^\star_{s,\mathrm{ph},1/2}$ *given by (6.3) for* $\delta = 1/2$, *and for all* $s > 3/2$, *the observability inequality (2.9) holds uniformly as* $h \to 0$.

*Proof (of Theorem 6.4).* The proof of our main result follows a dyadic decomposition argument that we explained in detail in Sect. 6.2 and we adapted to the case of data given by the null jump condition (6.67) and preconditioned by the Fourier truncation method in Sect. 6.3. Here, we only emphasize the main differences with respect to the previous two similar proofs. Thus, **Step I** is subdivided into two parts. The first one contains the new auxiliary result below, stating the fact that the total energy of data satisfying both hypotheses of Theorem 6.4 can be bounded by the energy concentrated on the low-frequency half of both branches of the spectrum.

**Proposition 6.3.** *For all initial data* $\mathbf{u^{h,\mathit{i}}} = (\{\mathbf{u}^{h,i}\}, [\mathbf{u}^{h,i}])$ *in (2.6) and (2.7) satisfying both conditions (6.67) and the fact that the average components* $\{\mathbf{u}^{h,i}\}$ *belong to the bi-grid class* $\mathscr{B}^h$ *in (6.30) for all* $i = 0, 1$, *and for all* $s > 1$, *the following estimate holds:*

$$\mathscr{E}_s^{\varsigma,h}(\mathbf{u^{h,0}}, \mathbf{u^{h,1}}) \leq 2\mathscr{E}_s^{\varsigma,h}(\Gamma_{\Pi^{2h}}\mathbf{u^{h,0}}, \Gamma_{\Pi^{2h}}\mathbf{u^{h,1}}), \tag{6.89}$$

*where* $\varsigma \in \{W, K\}$ *and* $\Gamma_{\Pi^{2h}}$ *is the projection defined by (6.31), with* $\Pi^{h,\sharp} = \Pi^{2h} := [-\pi/2h, \pi/2h]$.

We give the proof of Proposition 6.3 after the one of Theorem 6.4.

**Step I** consists in applying Proposition 6.2 with $\delta = 1/2$, so that

$$\mathscr{E}_s^{\varsigma,h}(\Gamma_{\Pi^{2h}}\mathbf{u}^{\mathbf{h},0}, \Gamma_{\Pi^{2h}}\mathbf{u}^{\mathbf{h},1}) \leq C(1/2)\mathscr{E}_s^{\varsigma,h}(\Gamma_{\text{ph}}\Gamma_{\Pi^{2h}}\mathbf{u}^{\mathbf{h},0}, \Gamma_{\text{ph}}\Gamma_{\Pi^{2h}}\mathbf{u}^{\mathbf{h},1}), \tag{6.90}$$

where $\varsigma \in \{W, K\}$, $C(1/2) > 0$ is the constant in (6.78) with $\delta = 1/2$ and $\Gamma_{\text{ph}}$ is the projection on the physical mode defined by (6.74). In **Step II**, the projectors $\wp_k$ are constructed so that both conditions (6.34) and (6.79) are satisfied simultaneously with $\delta = 1/2$. The rest of the proof follows the methodology described in Sects. 6.2 and 6.3. □

*Proof (of Proposition 6.3).* We consider only the case of $\varsigma = W$, for $\varsigma = K$ the arguments being similar. By applying Lemma 6.3 to the average part of the initial data $\{\mathbf{u}^{h,i}\}$, which belong to the bi-grid class $\mathscr{B}^h$ in (6.30), we have

$$\hat{u}^{h,\{\cdot\},i}(\xi) = \cos^2\left(\frac{\xi h}{2}\right)\hat{u}^{2h,\{\cdot\},i}(\xi) \quad \forall \xi \in \Pi^h \text{ and } i = 0, 1, \tag{6.91}$$

where $\hat{u}^{2h,\{\cdot\},i}(\xi)$ is the SDFT at the scale $2h$ of the sequence $(\{u^{h,i}\}(x_{2j+1}))_{j\in\mathbb{Z}}$. Taking into account that the initial data $\mathbf{u}^{\mathbf{h},i}$ in (2.6) satisfy the null jump condition (6.67) for both $i = 0, 1$, their total energy is given by (6.68). Then, using (6.91) and the $\pi/h$-periodicity of $\hat{u}^{2h,\{\cdot\},i}(\xi)$ in the identity (6.60), we get

$$\begin{aligned}&\mathscr{E}_s^{W,h}((\Gamma_{\Pi^h} - \Gamma_{\Pi^{2h}})\mathbf{u}^{\mathbf{h},0}, (\Gamma_{\Pi^h} - \Gamma_{\Pi^{2h}})\mathbf{u}^{\mathbf{h},1})\\ &= \frac{1}{4\pi}\int_{\Pi^{2h}}\left[\frac{2+\cos(\xi h)}{3}\hat{m}^{h,\flat}(\xi)|\hat{u}^{h,\{\cdot\},1}(\xi)|^2 + \hat{r}^{h,\flat}(\xi)\hat{\Lambda}^h_{1,\text{ph}}(\xi)|\hat{u}^{h,\{\cdot\},0}(\xi)|^2\right]d\xi,\end{aligned} \tag{6.92}$$

where

$$\hat{m}^{h,\flat}(\xi) := \frac{2-\cos(\xi h)}{2+\cos(\xi h)}\tan^4\left(\frac{\xi h}{4}\right) \text{ and } \hat{r}^{h,\flat}(\xi) := \tan^2\left(\frac{\xi h}{2}\right).$$

An easy computation shows that both $\hat{m}^{h,\flat}(\xi)$ and $\hat{r}^{h,\flat}(\xi)$ are increasing functions on $\Pi^{2h}$, so that their maximum value is attained at $\xi = \pi/2h$ and $\|\hat{m}^{h,\flat}\|_{L^\infty(\Pi^{2h})} = \|\hat{r}^{h,\flat}\|_{L^\infty(\Pi^{2h})} = 1$. Therefore,

$$\mathscr{E}_s^{W,h}((\Gamma_{\Pi^h} - \Gamma_{\Pi^{2h}})\mathbf{u}^{\mathbf{h},0}, (\Gamma_{\Pi^h} - \Gamma_{\Pi^{2h}})\mathbf{u}^{\mathbf{h},1}) \leq \mathscr{E}_s^{W,h}(\Gamma_{\Pi^{2h}}\mathbf{u}^{\mathbf{h},0}, \Gamma_{\Pi^{2h}}\mathbf{u}^{\mathbf{h},1}),$$

which, in view of (6.60), concludes the proof of Proposition 6.3. □

## 6.5 Numerical Experiments

In this section, the initial data in (2.6) are constructed out of the following Gaussian profile in the Fourier space:

$$\hat{\sigma}_\gamma(\xi) = \sqrt{\frac{2\pi}{\gamma^2}} \exp\left(-\frac{\xi^2}{2\gamma^2}\right), \text{ with } \gamma = h^{-1/2}. \tag{6.93}$$

Our aim is to highlight both the pathological phenomena presented in Theorem 5.1 of Chap. 5 and the effect of the two bi-grid filtering algorithms in Sects. 6.2 and 6.4 on these high-frequency Gaussian wave packets. Let us remark that the value of $\gamma := h^{-1/2}$ we consider in these numerical simulations fulfills the requirements (5.9). However, as emphasized in [49], this value of $\gamma$ is the critical one separating the region in which only the transport effect of the group velocity is dominant (for $\gamma \leq h^{-1/2}$) and the one in which higher-order effects of the dispersion relation destroy the initial Gaussian shape of the wave packets as time evolves (for $\gamma >> h^{-1/2}$).

All the numerical simulations in this section are done taking the penalty parameter of the SIPG method to be $s = 5$, the mesh size parameter $h = 1/10000$, and the final time $t = 1$.

We address the following four situations:

**I. Data concentrated on the physical mode** (i.e., satisfying (5.1)) such that

$$\hat{u}^{h,1}(\xi) = i\hat{\lambda}^h_{s,\mathrm{ph}}(\xi)\hat{u}^{h,0}(\xi) \text{ and } \hat{u}^{h,0}(\xi) := \hat{\sigma}_\gamma(\xi - \xi_0)\chi_{\Pi^h}(\xi),$$

with $\hat{\sigma}_\gamma$ as in (6.93) and $\xi_0 = \eta_0/h \in \Pi^h$. Then the solution of (2.13) is given by (5.6), so that both average and jump components are wave packets propagating to the left at the same group velocity $\partial_\xi\hat{\lambda}^h_{s,\mathrm{ph}}(\xi_0)$. Observe that this group velocity is larger than the continuous one, $\partial_\xi\hat{\lambda}(\xi) \equiv 1$ for $\xi_0 = 21\pi/32h$ (see Fig. 6.1), while for $\xi_0 = 19\pi/20h$ it is smaller than the characteristic group velocity (see Fig. 6.2 top left).

**II. Bi-Grid data concentrated on the physical mode**, i.e., satisfying (5.1) and such that $\hat{u}^{h,0}(\xi)$ and $\hat{u}^{h,1}(\xi)$ are the bi-grid projections of the Gaussian profiles $\hat{\vartheta}^{h,0}(\xi) := \hat{\sigma}_\gamma(\xi - \xi_0)\chi_{\Pi^h}(\xi)$ and $\hat{\vartheta}^{h,1}(\xi) := i\hat{\lambda}^h_{s,\mathrm{ph}}(\xi)\hat{\vartheta}^{h,0}(\xi)$ [with $\hat{\sigma}_\gamma$ as in (6.93)]. The procedure to apply the bi-grid filtering on Gaussian initial data is described in detail in [51] in the context of the linear Schrödinger equation. Roughly speaking, it can be done in two steps:

1. The obtention of a new Gaussian-type profile on $\Pi^{2h}$ produced by adding the restrictions of $\hat{\phi}^{h,0}(\xi)$ and $\hat{\phi}^{h,1}(\xi)$ to $[-\pi/h, -\pi/2h]$ and $[\pi/2h, \pi/h]$ to their values on $[0, \pi/2h]$ and $[-\pi/2h, 0]$;

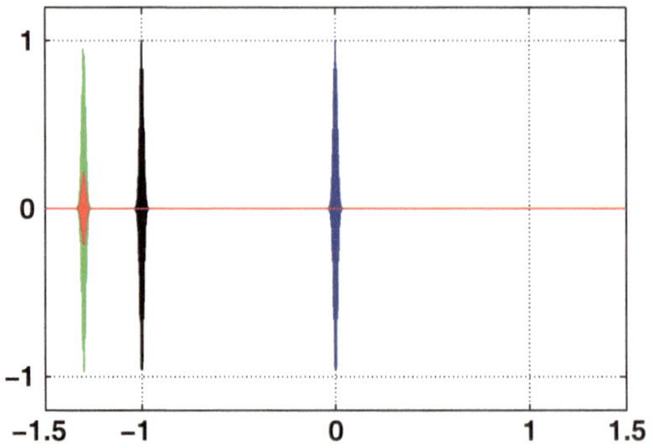


**Fig. 6.1** Average/jump components (in *green/red*) of the solution to the SIPG approximation of the wave Eq. (1.18) at time $t = 1$ consisting in a Gaussian wave packet concentrated on the physical mode at the wave number $\xi_0 = 21\pi/32h$ without any filtering on $(\mathbf{u}^{h,0}, \mathbf{u}^{h,1})$ in (5.1) compared to the initial data (*blue*) and the solution of the continuous model at time $t = 1$ (*black*)

2. Extension of the new profiles on $\Pi^{2h}$ by $\pi/h$-periodicity to $\Pi^h$ and multiplication of the results by the Fourier symbol of the bi-grid algorithm of mesh ratio $1/2$, $\hat{b}^h(\xi) := \cos^2(\xi h/2)$.

In the physical space, these two steps correspond to

1. Producing data on the coarse grid by restricting the data on the fine grid to the odd (or even) grid points
2. Extending the data on the coarse grid by linear interpolation to the fine one

More precisely, the bi-grid data we consider are approximately

$$\hat{u}^{h,0}(\xi) \sim (\hat{\sigma}_\gamma(\xi - \xi_0) + \hat{\sigma}_\gamma(\xi + \pi/h - \xi_0))\chi_{\Pi^h}(\xi)\hat{b}^h(\xi)$$

and

$$\hat{u}^{h,1}(\xi) \sim i\hat{\lambda}^h_{s,\mathrm{ph}}(\xi_0)(\hat{\sigma}_\gamma(\xi - \xi_0) + \hat{\sigma}_\gamma(\xi + \pi/h - \xi_0))\chi_{\Pi^h}(\xi)\hat{b}^h(\xi).$$

For these initial data, the solution (5.3) of (2.13) takes the particular form

$$\hat{\mathbf{u}}^h_s(\xi,t) \sim \hat{\mathbf{v}}^h_{s,\mathrm{ph}}(\xi)\hat{b}^h(\xi)\chi_{\Pi^h}(\xi)\sum_{\pm}\Bigg[\left(\frac{1}{2} \pm \frac{1}{2}\right)\hat{\sigma}_\gamma(\xi - \xi_0)$$
$$+\frac{1}{2}\left(1 \pm \frac{\hat{\lambda}^h_{s,\mathrm{ph}}(\xi_0)}{\hat{\lambda}^h_{s,\mathrm{ph}}(\xi)}\right)\hat{\sigma}_\gamma(\xi + \pi/h - \xi_0)\Bigg]\exp(\pm it\hat{\lambda}^h_{s,\mathrm{ph}}(\xi)). \quad (6.94)$$

In this way, both average and jump components split into three wave packets, one of them propagating to the left at the group velocity $\partial_\xi\hat{\lambda}^h_{s,\mathrm{ph}}(\xi_0)$ and two

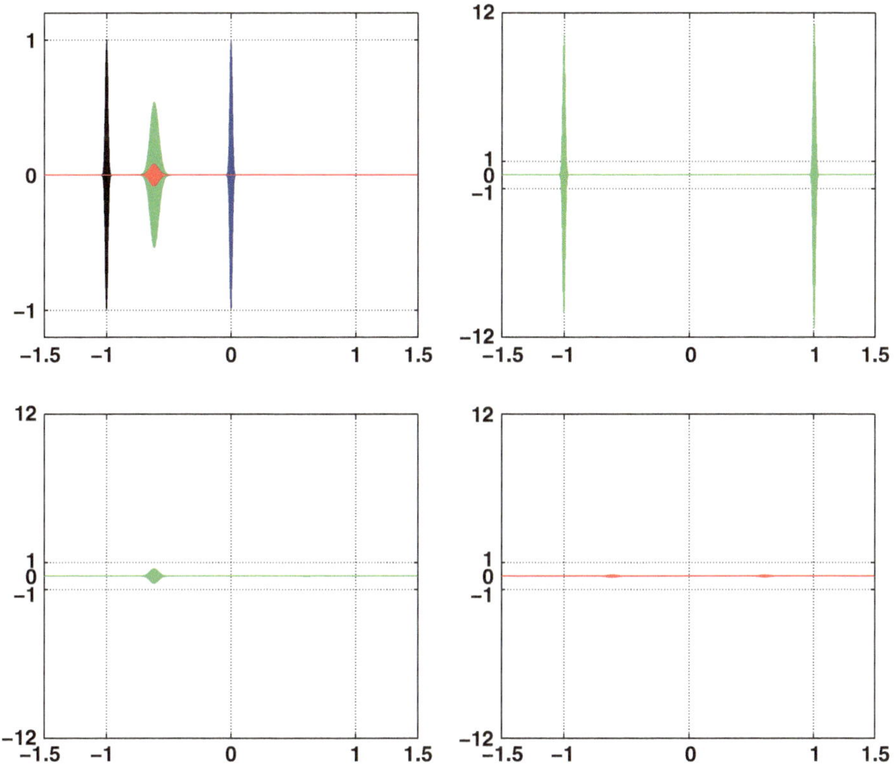


**Fig. 6.2** Average/jump components (in green/red) of the solution for the SIPG approximation of the wave equation at the wave number $\xi_0 = 19\pi/20h$ and at $t = 1$ compared to the initial data (*blue*) and the solution of the wave Eq. (1.18) at time $t = 1$ (*black*). *Top, left:* Averages and jumps of the solution concentrated on the physical mode and without any filtering on $(\mathbf{u}^{h,0}, \mathbf{u}^{h,1})$ in (5.1). *Top, right:* Averages of the solution concentrated on the physical mode with bi-grid filtering on $(\mathbf{u}^{h,0}, \mathbf{u}^{h,1})$. *Bottom, left and right:* Averages and jumps of the solution corresponding to data with null jumps and no filtering on the averages $(\{\mathbf{u}^{h,0}\}, \{\mathbf{u}^{h,1}\})$

of them propagating in both directions with the velocity $\partial_\xi \hat{\lambda}^h_{s,\mathrm{ph}}(\xi_0 - \pi/h)$ (see Fig. 6.3 top left and right when $\xi_0 = 21\pi/32h$). For $\xi_0 = 19\pi/20h$, the wave packet propagating at velocity $\partial_\xi \hat{\lambda}^h_{s,\mathrm{ph}}(\xi_0)$ disappears for both average and jump components due to the effect of the multiplicative symbol $\hat{b}^h(\xi) := \cos^2(\xi h/2)$ vanishing at $\pi/h$ and then having small amplitude for wave numbers $\xi_0$ close to $\pi/h$ (see Fig. 6.2 top right). The two remaining wave packets have large amplitudes due to the factor

$$\frac{1}{2}\left(1 \pm \frac{\hat{\lambda}^1_{s,\mathrm{ph}}(\eta_0)}{\hat{\lambda}^1_{s,\mathrm{ph}}(\eta_0 - \pi)}\right),$$

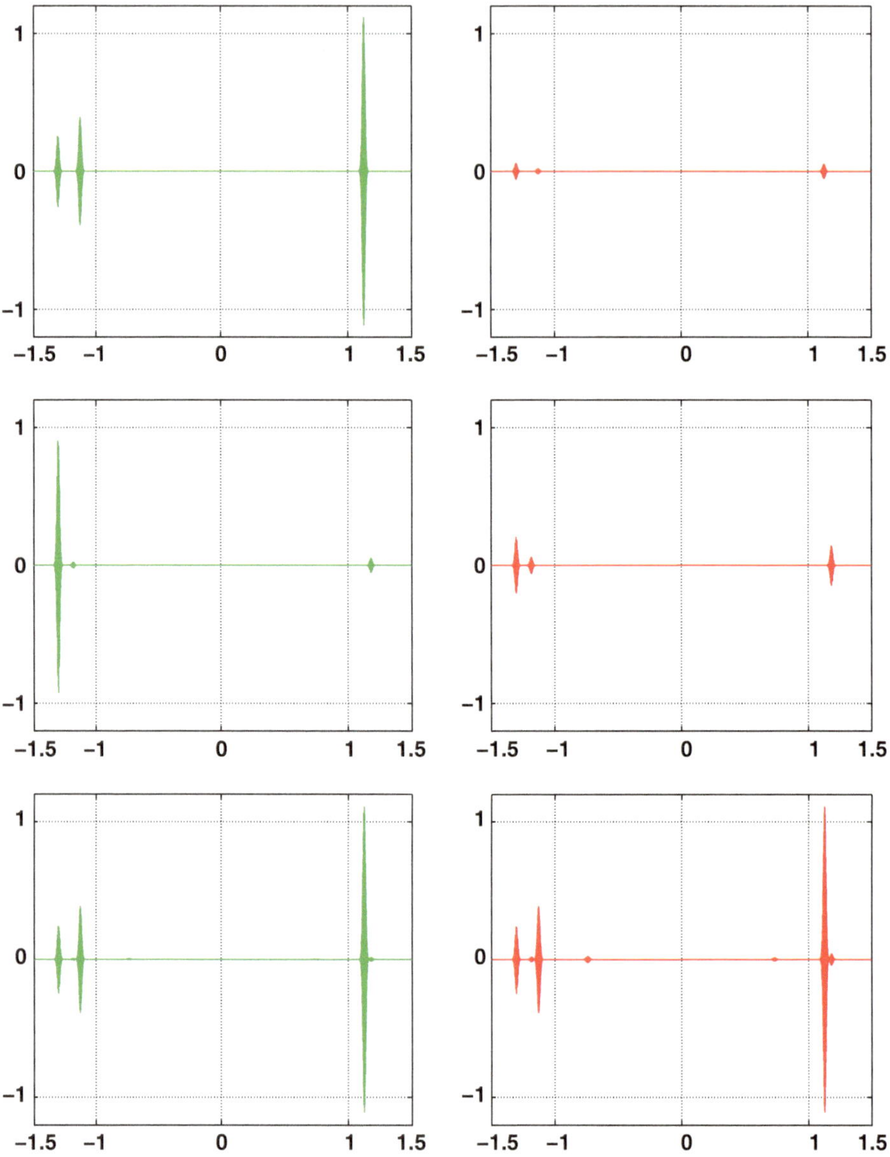


**Fig. 6.3** Average/jump components (in *green/red*) of the solution to the SIPG approximation of the wave Eq. (1.18) at the wave number $\xi_0 = 21\pi/32h$ and at time $t = 1$. *Top, left and right:* Averages and jumps of the solution concentrated on the physical mode with bi-grid filtering on $(\mathbf{u}^{h,0}, \mathbf{u}^{h,1})$ in (5.1). *Middle, left and right:* Averages and jumps of the solution corresponding to data with null jumps and no filtering on the averages $(\{\mathbf{u}^{h,0}\}, \{\mathbf{u}^{h,1}\})$. *Bottom, left and right:* Averages and jumps of the solution corresponding to data with null jumps and bi-grid filtering on the averages $(\{\mathbf{u}^{h,0}\}, \{\mathbf{u}^{h,1}\})$

which is large when $\eta_0$ is close to $\pi$ due to the fact that $\hat{\lambda}^1_{s,\mathrm{ph}}(\xi) \sim 0$ close to the origin $\xi = 0$. For the corresponding jump component (that we have not represented), the wave packet propagating at velocity $\partial_\xi \hat{\lambda}^h_{s,\mathrm{ph}}(\xi_0 - \pi/h)$ also disappears due to the fact that $h\xi_0 - \pi = -1/50 \sim 0$ and due to the behavior of the physical eigenvector $\hat{\mathbf{v}}^h_{s,\mathrm{ph}}(\xi)$ which is one of the factors of the solution (5.6) and whose second component vanishes as $\xi = 0$ (cf. property (a1) in Proposition 4.1).

**III. Data with null jump components** such that $\hat{u}^{h,\{\cdot\},1}(\xi) = i\hat{\lambda}^h_{s,\mathrm{ph}}(\xi)\hat{u}^{h,\{\cdot\},0}(\xi)$ and $\hat{u}^{h,\{\cdot\},0}(\xi) := \hat{\sigma}_\gamma(\xi-\xi_0)\chi_{\Pi^h}(\xi)$, with $\hat{\sigma}_\gamma$ as in (6.93). In this case, the components $\hat{u}^h_{s,\mathrm{ph}}(\xi,t)$ and $\hat{u}^h_{s,\mathrm{sp}}(\xi,t)$ entering in the solution (6.69) of (2.13) can be simplified as follows:

$$\hat{u}^h_{s,\mathrm{ph}}(\xi,t) = \hat{\sigma}_\gamma(\xi-\xi_0)\chi_{\Pi^h}(\xi)\exp\big(it\hat{\lambda}^h_{s,\mathrm{ph}}(\xi)\big) \tag{6.95}$$

and

$$\hat{u}^h_{s,\mathrm{sp}}(\xi,t) = \hat{\sigma}_\gamma(\xi-\xi_0)\chi_{\Pi^h}(\xi)\sum_{\pm}\frac{1}{2}\left(1 \pm \frac{\hat{\lambda}^h_{s,\mathrm{ph}}(\xi)}{\hat{\lambda}^h_{s,\mathrm{sp}}(\xi)}\right)\exp\big(\pm it\hat{\lambda}^h_{s,\mathrm{sp}}(\xi)\big).$$

Thus, in principle, both average and jump components of the numerical solution split in three wave packets. One of them corresponds to $\hat{u}^h_{s,\mathrm{ph}}(\xi,t)$ (which involves only the complex exponential $\exp(it\hat{\lambda}^h_{s,\mathrm{ph}}(\xi))$) and propagates with velocity $\partial_\xi\hat{\lambda}^h_{s,\mathrm{ph}}(\xi_0)$ in the negative direction. The remaining two wave packets correspond to $\hat{u}^h_{s,\mathrm{sp}}(\xi,t)$ involving both exponentials $\exp\big(\pm it\hat{\lambda}^h_{s,\mathrm{sp}}(\xi)\big)$ and propagating with velocity $\partial_\xi\hat{\lambda}^h_{s,\mathrm{sp}}(\xi_0)$ in both positive and negative directions (see Fig. 6.3 middle left and right for $\xi_0 = 21\pi/32h$). For $\xi_0 = 19\pi/20h$ ($\sim \pi/h$), the component $\hat{u}^h_{s,\mathrm{sp}}(\xi,t)$ has small amplitude (see Fig. 6.2 bottom left and right) due to the behavior of the matrix $\hat{\mathbf{V}}^{h,\sharp}_s(\xi)$ in (6.69) at $\xi = \pi/h$ (its component $(1,1)$ is one and the other ones vanish).

**IV. Bi-grid data with null jump components** are such that $\hat{u}^{h,\{\cdot\},0}(\xi)$ and $\hat{u}^{h,\{\cdot\},1}(\xi)$ in (6.69) are the bi-grid projections of the Gaussian profiles $\hat{\vartheta}^{h,\{\cdot\},0}(\xi) := \hat{\sigma}_\gamma(\xi-\xi_0)\chi_{\Pi^h}(\xi)$ and $\hat{\vartheta}^{h,\{\cdot\},1}(\xi) = i\hat{\lambda}^h_{s,\mathrm{ph}}(\xi)\hat{\vartheta}^{h,\{\cdot\},0}(\xi)$ (with $\widehat{\sigma}_\gamma$ as in (6.93)). In this case, both averages and jumps of the numerical solution split into seven wave packets. Three of them correspond to $\hat{u}^h_{s,\mathrm{ph}}(\xi,t)$ in (6.95) and propagate at velocity $\partial_\xi\hat{\lambda}^h_{s,\mathrm{ph}}(\xi_0)$ in the negative direction and at speed $\partial_\xi\hat{\lambda}^h_{s,\mathrm{ph}}(\xi_0-\pi/h)$ in both positive and the negative directions. The remaining four wave packets correspond to $\hat{u}^h_{s,\mathrm{sp}}(\xi,t)$ and propagate at velocities $\partial_\xi\hat{\lambda}^h_{s,\mathrm{sp}}(\xi_0)$ and $\partial_\xi\hat{\lambda}^h_{s,\mathrm{sp}}(\xi_0-\pi/h)$ in both negative and positive directions (see Fig. 6.3 bottom left and right).

# Chapter 7
# Extensions to Other Numerical Approximation Schemes

This chapter is aimed to show that the SIPG method is not the unique one having the pathologies under consideration and for which the filtering mechanisms here designed are suitable. In fact, these are typical situations for both *classical* and *nonconforming* methods, as we will see from the three examples in this chapter: a classical FEM (the quadratic one) and two more $P_1$-DG methods (the local DG method a version of the SIPG method penalizing the jumps across the grid points of both the numerical solution and its normal derivative).

## 7.1 The Quadratic Classical Finite Element Method ($P_2$-FEM)

Denote by $\mathscr{P}_2^h$ the space of *piecewise quadratic functions* on the uniform grid $\mathscr{G}^h$ and by $\mathscr{V}_q^h := H^1(\mathbb{R}) \cap \mathscr{P}_2^h$ the space of *piecewise quadratic* and *continuous finite elements* to be considered in this section (the subscript $q$ standing for *quadratic*). This space is generated by two kinds of basis functions, $\phi_j(x)$ and $\phi_{j+1/2}(x)$, interpolating the values of the numerical solution at the *nodal points* $x_j$ and at the *midpoints* $x_{j+1/2}$, with $j \in \mathbb{Z}$ (see Fig. 7.1).

Let us introduce the corresponding bilinear form given by

$$\mathscr{A}^h : \mathscr{V}_q^h \times \mathscr{V}_q^h \to \mathbb{R}, \quad \mathscr{A}^h(u^h, v^h) := (\partial_x^h u^h, \partial_x^h v^h)_{L^2(\mathscr{T}^h)}$$

and, associated to it, the $P_2$-FEM semi-discretization of the wave equation, which can be written in a variational form similar to (2.2), but with the space $\mathscr{V}^h$, the bilinear form $\mathscr{A}_s^h$, and the solution $u_s^h(\cdot, t)$ being replaced by $\mathscr{V}_q^h$, $\mathscr{A}^h$ and

$$u_q^h(x,t) = \sum_{j\in\mathbb{Z}} (u_j(t)\phi_j(x) + u_{j+1/2}(t)\phi_{j+1/2}(x)).$$

A. Marica and E. Zuazua, *Symmetric Discontinuous Galerkin Methods for 1 – D Waves*, SpringerBriefs in Mathematics, DOI 10.1007/978-1-4614-5811-1_7, 

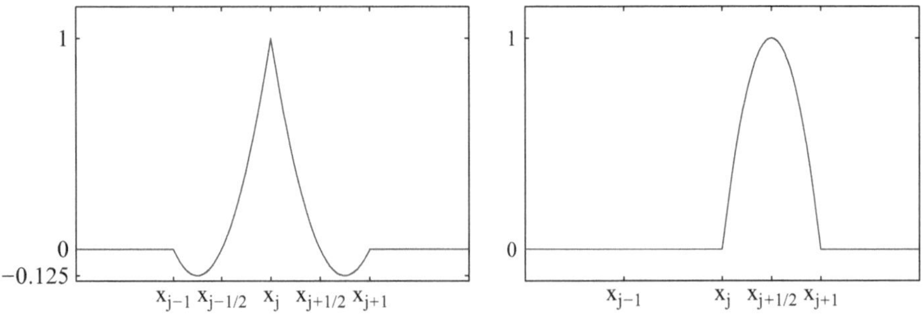


**Fig. 7.1** Basis functions for the $P_2$-FEM method: $\phi_j$ (*left*) and $\phi_{j+1/2}$ (*right*)

Set $\mathbf{u}_q^{\mathbf{h}}(t) := (u_j(t), u_{j+1/2}(t))_{j\in\mathbb{Z}}$. The $P_2$-FEM semi-discretization of the $1-d$ wave equation can be written in the matrix form (2.6), in which the sequence $\mathbf{u}_s^{\mathbf{h}}(t)$ is replaced by $\mathbf{u}_q^{\mathbf{h}}(t)$ and the *mass* and *stiffness* matrices $\mathbf{M}^{\mathbf{h}}$ and $\mathbf{R}_s^{\mathbf{h}}$ by the *pentadiagonal* ones, $\mathbf{M}_q^{\mathbf{h}}$ and $\mathbf{R}_q^{\mathbf{h}}$, generated by the stencils:

$$M_q^h = \begin{pmatrix} -\frac{h}{30} & \frac{h}{15} & \frac{4h}{15} & \frac{h}{15} & -\frac{h}{30} \\ 0 & 0 & \frac{h}{15} & \frac{8h}{15} & \frac{h}{15} \end{pmatrix} \text{ and } R_q^h = \begin{pmatrix} \frac{1}{3h} & -\frac{8}{3h} & \frac{14}{3h} & -\frac{8}{3h} & \frac{1}{3h} \\ 0 & 0 & -\frac{8}{3h} & \frac{16}{3h} & -\frac{8}{3h} \end{pmatrix}. \quad (7.1)$$

By applying the SDFT to system (2.6) (with the above changes), we obtain a second-order ODE system similar to (2.13), in which the two Fourier symbols $\hat{\mathbf{M}}^h(\xi)$ and $\hat{\mathbf{R}}_s^h(\xi)$ are replaced by $\hat{\mathbf{M}}_q^h(\xi)$ and $\hat{\mathbf{R}}_q^h(\xi)$ below:

$$\hat{\mathbf{M}}_q^h(\xi) = \begin{pmatrix} \frac{4-\cos(\xi h)}{15} & \frac{2}{15}\cos\left(\frac{\xi h}{2}\right) \\ \frac{2}{15}\cos\left(\frac{\xi h}{2}\right) & \frac{8}{15} \end{pmatrix}, \quad \hat{\mathbf{R}}_q^h(\xi) = \frac{1}{h^2}\begin{pmatrix} \frac{14+2\cos(\xi h)}{3} & -\frac{16}{3}\cos\left(\frac{\xi h}{2}\right) \\ -\frac{16}{3}\cos\left(\frac{\xi h}{2}\right) & \frac{16}{3} \end{pmatrix}.$$

The so-called *acoustic* (physical) and *optic* (spurious) Fourier symbols, $\hat{\Lambda}_a^h(\xi)$ and $\hat{\Lambda}_o^h(\xi)$, are given by

$$\hat{\Lambda}_\alpha^h(\xi) := \frac{1}{h^2}\frac{22 + 8\cos^2\left(\frac{\xi h}{2}\right) + 2\text{sign}(\alpha)\sqrt{\hat{\Delta}^h(\xi)}}{1 + \sin^2\left(\frac{\xi h}{2}\right)},$$

where $\alpha \in \{a, o\}$, $\text{sign}(a) = -1$, $\text{sign}(o) = 1$, and

$$\hat{\Delta}^h(\xi) := 1 + 268\cos^2\left(\frac{\xi h}{2}\right) - 44\cos^4\left(\frac{\xi h}{2}\right).$$

It can be easily verified that the two *dispersion relations* in (2.17), $\hat{\lambda}_a^h(\xi)$ and $\hat{\lambda}_o^h(\xi)$, have a similar configuration to the one for $s = 5$ in Fig. 2.2, i.e., the *acoustic/optic* branch is an *increasing/decreasing* function of $\xi \in [0, \pi/h]$. The two Fourier modes

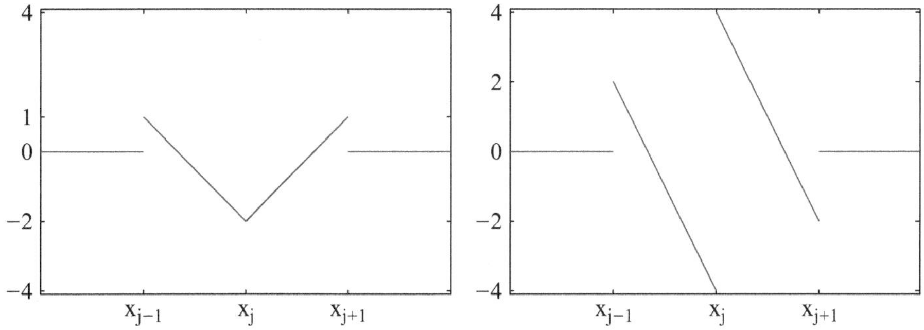


**Fig. 7.2** Functions $\phi_j^{\sharp}(x)$ (*left*) and $\phi_j^{\natural}(x)$ (*right*) generating the lifting operators $\mathscr{R}([\cdot])$ and $\mathscr{L}([\cdot])$

are well separated and $\hat{\lambda}_a^h(\pi/h) = \sqrt{10}/h$, $\hat{\lambda}_o^h(\pi/h) = \sqrt{12}/h$ and $\hat{\lambda}_o^h(0) = \sqrt{60}/h$. Moreover, the *acoustic group velocity* $\partial_\xi \hat{\lambda}_a^h(\xi)$ vanishes at $\xi = \pi/h$, while the *optic* one $\partial_\xi \hat{\lambda}_o^h(\xi)$ vanishes at both $\xi = 0$ and $\xi = \pi/h$.

Consequently, the observability inequality (2.9) corresponding to the $P_2$-FEM approximation holds nonuniformly as the mesh size parameter tends to zero and the filtering mechanisms in Chap. 6 have similar statements for the numerical approximation under consideration. For example, the filtering strategy in Sect. 6.4 could be adapted so that *linear data* with *nodal components given by a bi-grid filtering mechanism* are required. For further details, see [52] and [53] for the $P_2$-FEM approximation of the *exact boundary controllability* problem for the $1-d$ wave equation.

## 7.2 Local Discontinuous Galerkin (LDG) Methods

For all $\phi^h \in \mathscr{V}^h$, we introduce the *lifting operators* $\mathscr{R}, \mathscr{L} : \ell^2(\mathscr{G}^h) \to \mathscr{V}^h$ defined as

$$(\mathscr{R}(\varphi^h), \phi^h)_{L^2} = -(\varphi^h, \{\boldsymbol{\phi}^h\})_{\ell^2(\mathscr{G}^h)}, \; (\mathscr{L}(\varphi^h), \phi^h)_{L^2} = -(\varphi^h, [\boldsymbol{\phi}^h])_{\ell^2(\mathscr{G}^h)}. \quad (7.2)$$

Here, $L^2 := L^2(\mathbb{R})$. The following result provides a complete characterization of these lifting operators acting on the jump component, $\mathscr{R}([\cdot])$ and $\mathscr{L}([\cdot])$:

**Proposition 7.1.** *For all $f^h \in \mathscr{V}^h$, the lifting operators $\mathscr{R}([\mathbf{f}^h]), \mathscr{L}([\mathbf{f}^h]) \in \mathscr{V}^h$ have the explicit form:*

$$\mathscr{R}([\mathbf{f}^h])(x) = \frac{1}{h} \sum_{j \in \mathbb{Z}} \Big[ \Big( \frac{1}{2}[f^h](x_{j-1}) - 2[f^h](x_j) + \frac{1}{2}[f^h](x_{j+1}) \Big) \phi_j^{\{\cdot\}}(x)$$

$$+ ([f^h](x_{j-1}) - [f^h](x_{j+1})) \phi_j^{[\cdot]}(x) \Big]$$

*and*

$$\mathscr{L}([\mathbf{f}^h])(x) = \frac{1}{h}\sum_{j\in\mathbb{Z}}\Big[([f^h](x_{j+1}) - [f^h](x_{j-1}))\phi_j^{\{\cdot\}}(x)$$
$$- (2[f^h](x_{j+1}) + 8[f^h](x_j) + 2[f^h](x_{j-1}))\phi_j^{[\cdot]}(x)\Big].$$

Set $\phi_j^{\sharp}(x)$ and $\phi_j^{\natural}(x)$ to be the functions in Fig. 7.2. A simpler representation of $\mathscr{R}([\cdot])$ and $\mathscr{L}([\cdot])$ is

$$\mathscr{R}([\mathbf{f}^h])(x) = \frac{1}{h}\sum_{j\in\mathbb{Z}}[f^h](x_j)\phi_j^{\sharp}(x) \text{ and } \mathscr{L}([\mathbf{f}^h])(x) = \frac{1}{h}\sum_{j\in\mathbb{Z}}[f^h](x_j)\phi_j^{\natural}(x).$$

*Proof (of Proposition 7.1).* Our aim is to describe the way in which the lifting operators $\mathscr{R}$ and $\mathscr{L}$ act on the jump components of the elements in $\mathscr{V}^h$. We perform the analysis only for $\mathscr{R}$, the one of $\mathscr{L}$ being similar. Let us observe that for all $f^h \in \mathscr{V}^h$, $\mathscr{R}([\mathbf{f}^h])$ is also an element of $\mathscr{V}^h$, so that there exists a function $g^h \in \mathscr{V}^h$ of coefficients $\mathbf{g^h} = (\{g^h\}(x_j), [g^h](x_j))_{j\in\mathbb{Z}}$ so that $g^h = \mathscr{R}([\mathbf{f}^h])$. Following the definition of $\mathscr{R}$ in (7.2), we see that the sequence $\mathbf{g^h}$ is the solution of the system $\mathbf{M^h}\mathbf{g^h} = \mathbf{c^h}$, where the sequence $\mathbf{c^h} = (\{c^h\}(x_j), [c^h](x_j))_{j\in\mathbb{Z}}$ is defined by $\{c^h\}(x_j) := -[f^h](x_j)$ and $[c^h](x_j) := 0$. The inverse $(\mathbf{M^h})^{-1}$ of the mass matrix $\mathbf{M^h}$ is a block tridiagonal matrix generated by the stencil:

$$(M^h)^{-1} := \frac{1}{h}\left(\begin{array}{cc|cc|cc} -\frac{1}{2} & 1 & 2 & 0 & -\frac{1}{2} & -1 \\ -1 & 2 & 0 & 8 & 1 & 2 \end{array}\right). \tag{7.3}$$

This concludes the proof of Proposition 7.1. □

In what follows, we introduce the DG method under consideration in this section, the so-called *local discontinuous Galerkin* (LDG) method ([6, 20]), whose bilinear form $\mathscr{A}_{s,\beta}^h : \mathscr{V}^h \times \mathscr{V}^h \to \mathbb{R}$ below is defined for all $s > 0$ and $\beta \in \mathbb{R}$:

$$\mathscr{A}_{s,\beta}^h(u^h, v^h) := \mathscr{A}_s^h(u^h, v^h) + \mathscr{A}_{\sharp,\beta}^h(u^h, v^h),$$

where $\mathscr{A}_s^h$ is the bilinear form (2.1) associated to the SIPG method and $\mathscr{A}_{\sharp,\beta}^h$ is a form depending on the parameter $\beta$ given by

$$\mathscr{A}_{\sharp,\beta}^h(u^h, v^h) := \beta([\mathbf{u}^h], [\partial_x^h \mathbf{v}^h])_{\ell^2(\mathscr{G}^h)} + \beta([\partial_x^h \mathbf{u}^h], [\mathbf{v}^h])_{\ell^2(\mathscr{G}^h)}$$
$$+ (\mathscr{R}([\mathbf{u}^h]) + \beta\mathscr{L}([\mathbf{u}^h]), \mathscr{R}([\mathbf{v}^h]) + \beta\mathscr{L}([\mathbf{v}^h]))_{L^2(\mathbb{R})}.$$

We consider the LDG semi-discretization of the wave equation yielding the infinite system of second-order ODEs (2.6) in which the mass matrix $\mathbf{M^h}$ is the same for all the $P_1$-DG methods, while the stiffness matrix $\mathbf{R}_s^{\mathbf{h}}$ has to be replaced by the matrix $\mathbf{R}_{s,\beta}^{\mathbf{h}}$ generated by the stencil

$$R_{s,\beta}^h = \left( \begin{array}{cc|cc|cc} -\frac{1}{h} & -\frac{\beta}{h} & \frac{2}{h} & \frac{2\beta}{h} & -\frac{1}{h} & -\frac{\beta}{h} \\ -\frac{\beta}{h} & -\frac{1}{4h} + \frac{4\beta^2-1}{2h} & \frac{2\beta}{h} & \frac{2s-1}{2h} + \frac{2(4\beta^2+1)}{h} & -\frac{\beta}{h} & -\frac{1}{4h} + \frac{4\beta^2-1}{2h} \end{array} \right). \tag{7.4}$$

Let us denote by $\mathbf{u}_{s,\beta}^{\mathbf{h}}(t)$ the solution of the LDG semi-discretization of the wave equation. The change of the stiffness matrix modifies the Fourier symbol of the stiffness matrix $\hat{\mathbf{R}}_s^h(\xi)$ in (2.13), such that it has to be replaced by $\hat{\mathbf{R}}_{s,\beta}^h(\xi)$ below:

$$\hat{\mathbf{R}}_{s,\beta}^h(\xi) := \frac{1}{h^2} \begin{pmatrix} 4\sin^2\left(\frac{\xi h}{2}\right) & 4\beta\sin^2\left(\frac{\xi h}{2}\right) \\ 4\beta\sin^2\left(\frac{\xi h}{2}\right) & \hat{r}_{s,\beta}^h(\xi) \end{pmatrix},$$

where

$$\hat{r}_{s,\beta}^h(\xi) := s - \cos^2\left(\frac{\xi h}{2}\right) + 2 - \cos(\xi h) + 4\beta^2(2 + \cos(\xi h)).$$

The two eigenvalues of the matrix $\hat{\mathbf{S}}_{s,\beta}^h(\xi) := (\hat{\mathbf{M}}^h(\xi))^{-1}\hat{\mathbf{R}}_{s,\beta}^h(\xi)$ are

$$\hat{\Lambda}_{s,\beta,\alpha}^h(\xi) = \frac{1}{h^2}\Big[2\sin^2\left(\frac{\xi h}{2}\right)(2 - \cos(\xi h)) + 2(2 + \cos(\xi h))\hat{r}_{s,\beta}^h(\xi) \pm \sqrt{\hat{\Delta}_{s,\beta}^h(\xi)}\Big],$$

where $\alpha \in \{ph, sp\}$, $\text{sign}(ph) = -1$, and $\text{sign}(sp) = 1$, and

$$\begin{aligned} \hat{\Delta}_{s,\beta}^h(\xi) := &\Big[2\sin^2\left(\frac{\xi h}{2}\right)(2 - \cos(\xi h)) + 2(2 + \cos(\xi h))\hat{r}_{s,\beta}^h(\xi)\Big]^2 \\ &- 48\sin^2\left(\frac{\xi h}{2}\right)\Big[\hat{r}_{s,\beta}^h(\xi) - 4\beta^2\sin^2\left(\frac{\xi h}{2}\right)\Big]. \end{aligned}$$

Let us remark that $\hat{\Delta}_{s,\beta}^h(\xi) = 0$ if $\beta = s = 0$ and $\xi = 0$ or $\xi = \pi/h$, which can be proved in a similar way to Lemma 4.1. In what follows, we consider $s > 0$.

The two group velocities have the following explicit expressions:

$$\partial_\xi \hat{\lambda}_{s,\beta,\alpha}^h(\xi) = -\text{sign}(\alpha)\frac{\cos\left(\frac{\xi h}{2}\right)}{\sqrt{\hat{\Delta}_{s,\beta}^h(\xi)}} \frac{\sin\left(\frac{\xi h}{2}\right)}{h\hat{\lambda}_{s,\beta,\alpha}^h(\xi)} \hat{e}_{s,\beta,\alpha}^h(\xi),$$

where, for all $\alpha \in \{ph, sp\}$,

$$\hat{e}_{s,\beta,\alpha}^h(\xi) = \hat{f}_{s,\beta}^h(\xi)h^2\hat{\Lambda}_{s,\beta,\alpha}^h(\xi) + \hat{g}_{s,\beta}^h(\xi), \quad \hat{f}_{s,\beta}^h(\xi) := 2s + 2 + (16\beta^2 - 4)(2 + \cos(\xi h))$$

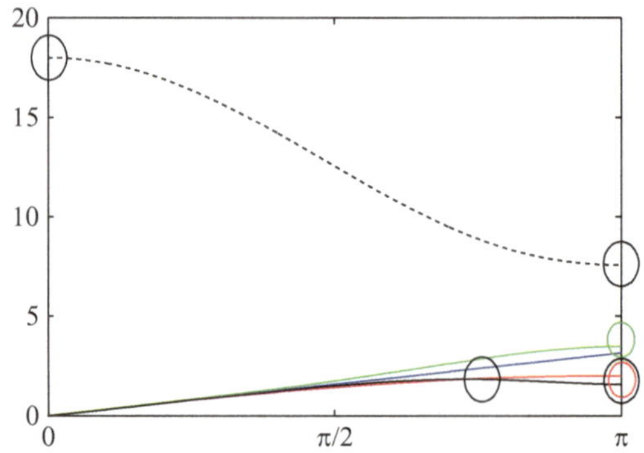


**Fig. 7.3** Physical/spurious dispersion relation for the LDG method, $\hat{\lambda}^1_{s,\beta,\alpha}(\xi)$, $\alpha = ph/sp$ (*solid/dotted black line*), compared to the ones of the continuous wave equation $\hat{\lambda}(\xi) = \xi$ (*blue*) and of its FD and $P_1$-FEM semi-discretizations, $\hat{\lambda}^1_{1,ph}(\xi)$ (*red*) and $\hat{\lambda}^1_{\infty,ph}(\xi)$ (*green*). The marked points are wave numbers where the corresponding group velocities vanish

and

$$\hat{g}^h_{s,\beta}(\xi) := 12\Big[\hat{r}^h_{s,\beta}(\xi) + (3 - 16\beta^2)\sin^2\Big(\frac{\xi h}{2}\Big)\Big].$$

For $s > 0$ and $\beta = 0$, it can be proved that $\hat{e}^h_{s,0,ph}(\xi) > 0$, such that the only critical point for the physical dispersion relation is $\xi = \pi/h$. The symbol $\hat{e}^h_{s,0,sp}(\xi)$ vanishes for some $\xi \in (0, \pi/h)$ for $s \in (0, 9/2)$. Then the spurious dispersion relation $\hat{\lambda}^h_{s,0,sp}(\xi)$ has three critical points (at $\xi = 0$, at $\xi = \pi/h$, and at an intermediate point $\xi \in (0, \pi/h)$), while, for $s \geq 9/2$, $\hat{e}^h_{s,0,sp}(\xi) > 0$ for all $\xi \in [0, \pi/h]$ and the spurious diagram $\hat{\lambda}^h_{s,0,sp}(\xi)$ is decreasing on $\xi \in [0, \pi/h]$ and has two critical points: $\xi = 0$ and $\xi = \pi/h$.

For $s > 0$ and $\beta \neq 0$, the situation is trickier. In particular, two critical points can appear on the physical dispersion relation: one, as usual, at $\xi = \pi/h$ and a second one on $(0, \pi/h)$.

If $\hat{f}^h_{s,\beta}(\xi) > 0$ and $\hat{g}^h_{s,\beta}(\xi) > 0$ hold simultaneously for all $\xi \in [0, \pi/h]$, one can guarantee that $\hat{e}^h_{s,\beta,ph}(\xi) > 0$ and that the only critical point of the physical dispersion relation is $\xi = \pi/h$. This holds, for example, when
(a) $\beta^2 < 1/4, s > 5 - 24\beta^2$ or (b) $\beta^2 > 1/4, s > \max\{1 - 8\beta^2, 12\beta^2 - 6\}$.

In Fig. 7.3, we consider the case ($s = 0.01, \beta = 1.5$) which does not fulfill either of the two conditions (a) or (b) above and for which the physical dispersion relation $\hat{\lambda}^h_{s,\beta,ph}(\xi)$ has two critical points.

However, whenever one of the two conditions (a) or (b) is satisfied, all the filtering techniques described in Chap. 6 work exactly as in the case of the SIPG method.

## 7.3 SIPG Methods with Penalization on the Normal Derivatives

In this section, we consider the so-called SIPG-n method, a version of the SIPG method which, apart from the jumps of the numerical solution across the grid points, also penalizes the jumps of its normal derivative (cf. [29]). The corresponding bilinear form $\mathscr{A}_{s,\beta}^{h,n} : \mathscr{V}^h \times \mathscr{V}^h \to \mathbb{R}$ below is defined for all $s > 1$ and $\beta > 0$:

$$\mathscr{A}_{s,\beta}^{h,n}(u^h, v^h) := \mathscr{A}_s^h(u^h, v^h) + \mathscr{A}_{\mathfrak{b},\beta}^h(u^h, v^h),$$

where $\mathscr{A}_s^h$ is the bilinear form (2.1) associated to the SIPG method and $\mathscr{A}_{\mathfrak{b},\beta}^h$ is a form depending on the parameter $\beta$ and given by

$$\mathscr{A}_{\mathfrak{b},\beta}^h(u^h, v^h) := \beta h([\partial_x^h \mathbf{u}^h], [\partial_x^h \mathbf{v}^h])_{\ell^2(\mathscr{G}^h)}.$$

We consider the SIPG-n semi-discretization of the wave equation (1.18) yielding the infinite system of second-order ODEs (2.6) in which the *mass* matrix $\mathbf{M^h}$ is maintained, while the *stiffness* matrix $\mathbf{R_s^h}$ is replaced by the *block pentadiagonal matrix* $\mathbf{R}_{s,\beta}^{\mathbf{h},n}$ generated by the stencil

$$R_{s,\beta}^{h,n} = \left(\begin{array}{cc|cc|cc|cc|cc} \frac{\beta}{h} & -\frac{\beta}{2h} & -\frac{1}{h} - \frac{4\beta}{h} & \frac{\beta}{h} & \frac{2}{h} + \frac{6\beta}{h} & 0 & -\frac{1}{h} - \frac{4\beta}{h} & -\frac{\beta}{h} & \frac{\beta}{h} & \frac{\beta}{2h} \\ \frac{\beta}{2h} & -\frac{\beta}{4h} & -\frac{\beta}{h} & -\frac{1}{4h} & 0 & \frac{\beta}{2h} + \frac{2s-1}{2h} & \frac{\beta}{h} & -\frac{1}{4h} & -\frac{\beta}{2h} & -\frac{\beta}{4h} \end{array}\right).$$

Let us denote by $\mathbf{u}_{s,\beta}^{\mathbf{h},n}(t)$ the solution of the SIPG-n approximation of the wave equation. The change of the stiffness matrix also modifies the corresponding Fourier symbol in (2.13), $\hat{\mathbf{R}}_s^h(\xi)$, which has to be replaced by $\hat{\mathbf{R}}_{s,\beta}^{h,n}(\xi)$ below:

$$\hat{\mathbf{R}}_{s,\beta}^{h,n}(\xi) := \frac{1}{h^2} \begin{pmatrix} 4\sin^2\left(\frac{\xi h}{2}\right) + 16\beta \sin^4\left(\frac{\xi h}{2}\right) & -4\beta i \sin(\xi h) \sin^2\left(\frac{\xi h}{2}\right) \\ 4\beta i \sin(\xi h) \sin^2\left(\frac{\xi h}{2}\right) & s - \cos^2\left(\frac{\xi h}{2}\right) + \beta \sin^2(\xi h) \end{pmatrix}.$$

The two eigenvalues of $\hat{\mathbf{S}}_{s,\beta}^{h,n}(\xi) := (\hat{\mathbf{M}}^h(\xi))^{-1} \hat{\mathbf{R}}_{s,\beta}^{h,n}(\xi)$ are

$$\hat{\Lambda}_{s,\beta,\alpha}^{h,n}(\xi) := \frac{1}{h^2}\left[12 + 2(s-3)(2 + \cos(\xi h)) + 24\beta \sin^2\left(\frac{\xi h}{2}\right) + \operatorname{sign}(\alpha)\sqrt{\hat{\Delta}_{s,\beta}^{h,n}(\xi)}\right],$$

where $\alpha \in \{ph, sp\}$, $\operatorname{sign}(ph) = -1$, $\operatorname{sign}(sp) = 1$, and

$$\hat{\Delta}_{s,\beta}^{h,n}(\xi) := \left[12 + 2(s-3)(2 + \cos(\xi h)) + 24\beta \sin^2\left(\frac{\xi h}{2}\right)\right]^2 - 48 \sin^2\left(\frac{\xi h}{2}\right)\left[s - \cos^2\left(\frac{\xi h}{2}\right)\right] - 192\beta s \sin^4\left(\frac{\xi h}{2}\right).$$

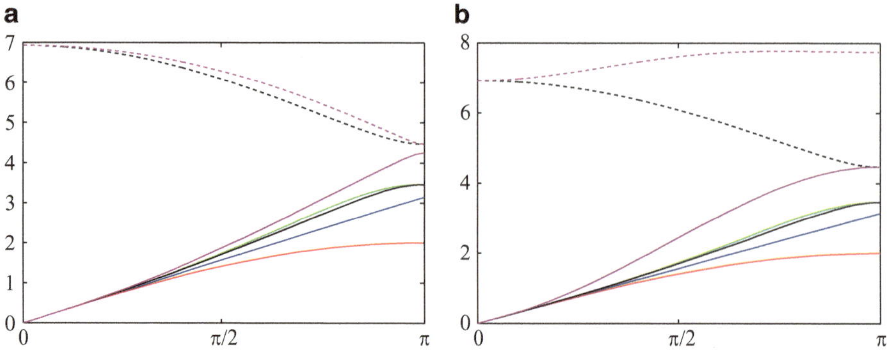


**Fig. 7.4** Dispersion relations for the SIPG method, $\hat{\lambda}^1_{s,\alpha}(\xi)$ (*solid/dotted black line* for $\alpha = ph/sp$) versus the ones for the SIPG-n method, $\hat{\lambda}^{1,n}_{s,\beta,\alpha}(\xi)$ (*magenta*), and those of the continuous wave equation $\hat{\lambda}(\xi) = \xi$ (*blue*) and of its FD and $P_1$-FEM schemes, $\hat{\lambda}^1_{1,ph}(\xi)$ (*red*) and $\hat{\lambda}^1_{\infty,ph}(\xi)$ (*green*). The two subfigures correspond to: (**a**) $s = 5$, $\beta = 0.125$ (**b**) $s = 5$, $\beta = 1$

Note that $\hat{\Delta}^{h,n}_{s,\beta}(\xi) = 0$ holds if and only if

- $\xi = 0$, $s = 1$, and $\beta \geq 0$ or
- $\xi = \pi/h$, $s = 3 + 12\beta$, and $\beta \geq 0$.

This can be proved in a similar way to Lemma 4.1, taking into account the fact that $\hat{\Delta}^{h,n}_{s,\beta}(\xi)$ can be written as a sum of two positive terms as

$$\hat{\Delta}^{h,n}_{s,\beta}(\xi) = \left[4\sin^2\left(\frac{\xi h}{2}\right)(1+6\beta) + 2(s-1)\left(3 - 4\sin^2\left(\frac{\xi h}{2}\right)\right)\right]^2 + 48\sin^2\left(\frac{\xi h}{2}\right)\cos^2\left(\frac{\xi h}{2}\right)(s-1)^2.$$

To ensure the stability of the numerical approximation, we take $s > 1$.

The two group velocities have the following explicit expressions:

$$\partial_\xi \hat{\lambda}^{h,n}_{s,\beta,\alpha}(\xi) = -\text{sign}(\alpha)\frac{\cos\left(\frac{\xi h}{2}\right)}{\sqrt{\hat{\Delta}^{h,n}_{s,\beta}(\xi)}}\frac{2\sin\left(\frac{\xi h}{2}\right)}{h\hat{\lambda}^{h,n}_{s,\beta,\alpha}(\xi)}\hat{e}^{h,n}_{s,\beta,\alpha}(\xi),$$

where, for all $\alpha \in \{ph, sp\}$,

$$\hat{e}^{h,n}_{s,\beta,\alpha}(\xi) = (s - 3 - 6\beta)h^2\hat{\Lambda}^{h,n}_{s,\beta,\alpha}(\xi) + 6(s - \cos(\xi h)) + 48\beta s\sin^2\left(\frac{\xi h}{2}\right).$$

It can be proved that, for all $s > 1$, $\beta \geq 0$, and $\xi \in [0, \pi/h]$, $\hat{e}^{h,n}_{s,\beta,ph}(\xi) > 0$, so that, similarly to the SIPG method (for which $\beta = 0$), the physical dispersion diagram $\hat{\lambda}^{h,n}_{s,\beta,ph}(\xi)$ is a strictly increasing function of $\xi \in [0, \pi/h]$ and it has a unique critical point at $\xi = \pi/h$, excepting the value $s = 3 + 12\beta$, when it has no critical point.

For $\beta \geq 0$ and $s \in ((5 + 36\beta + 48\beta^2)/(3 + 16\beta), 5/2 + 6\beta)$, $\hat{e}^{h,n}_{s,\beta,sp}(\xi)$ admits an unique root $\xi \in (0, \pi/h)$, such that the spurious diagram $\hat{\lambda}^{h,n}_{s,\beta,sp}(\xi)$ has changing monotonicity with three critical points: $\xi = 0$, $\xi = \pi/h$, and an intermediate wave number $\xi \in (0, \pi/h)$. For $s \in \big((1, \infty) \setminus ((5 + 36\beta + 48\beta^2)/(3 + 16\beta), 5/2 + 6\beta)\big) \setminus \{3 + 12\beta\}$, the spurious dispersion relation has two critical points, $\xi = 0$ and $\xi = \pi/h$. For $s = 3 + 12\beta$, it has only one critical point, $\xi = 0$ (see Fig. 7.4). Thus, all the filtering techniques described in Chap. 6 work similarly to the case of the SIPG method.

# Chapter 8
# Comments and Open Problems

DG methods provide a rich class of schemes allowing to modify the physical component of the classical finite element numerical approximations of PDEs by means of stabilization terms penalizing the jumps of the numerical solution in order to fit better the behavior of the continuous models. Despite of this, when approximating wave equations, they generate high-frequency wave packets propagating at an arbitrarily low velocity or even traveling in the wrong direction. We are able to design appropriate filtering mechanisms, more complex than for the FD or the $P_1$-FEM approximations, which allow us to recover the uniform observability properties for the DG schemes. Similar pathologies occur for higher-order classical approximations for the wave equation and in particular for the quadratic finite element methods (cf. [53]). Following [25], all the results in this book can be extended to any conservative fully discrete symmetric DG method for the wave equation.

In what follows, we list some interesting open subjects of research related to the problems, methodology, and results in this book. This list is not exhaustive. In particular, one could analyze:

- **Propagation properties of other DG methods**, like the *discontinuous Petrov–Galerkin* (DPG) ([14, 22]) or the *hybridizable DG* ones (cf. [18]). In particular, the DPG methods are known to be free of the so-called pollution effect, characterized by the degradation of the quality of the numerical solution when the frequency is increased, but the mesh size is maintained fixed (cf. [14]). The main difference of the DPG methods with respect to the SIPG ones is that the basis functions are not given a priori, but obtained through an optimization process. It would be interesting to analyze if this lack of pollution leads to the efficiency of the DPG methods to capturing the propagation/observability/control properties of the wave equation (1.18), uniformly with respect to the mesh size parameter.
- **DG approximations of the multidimensional wave equation and their propagation properties.** It is well known (cf. [3]) that, when dealing with tensor product finite elements and grids, the dispersion relations of the SIPG

A. Marica and E. Zuazua, *Symmetric Discontinuous Galerkin Methods for 1 – D Waves*, SpringerBriefs in Mathematics, DOI 10.1007/978-1-4614-5811-1_8,

approximation of the multidimensional wave equation can be written as a sum of one-dimensional dispersion relations, one for each coordinate direction. This does not happens for triangular multidimensional meshes or for DG methods obtained by writing the wave equation as a first-order PDE system in time and then discretizing the corresponding transport equations by DG methods for conservation laws. Consequently, developing the same kind of precise Fourier analysis as the one in this book in the multidimensional setting, even in the simple case of uniform grids and low-order polynomials, seems to be a more complex process.

- **Multiplier techniques for DG approximations of wave equations on bounded domains.** For the moment, there is no explicit spectral analysis of the DG approximations of the Laplace operator on bounded domains. Numerical computations show that the spectrum of the SIPG approximation of the Laplacian on a bounded interval is divided in two branches following basically the two Fourier symbols $\hat{\Lambda}^h_{s,ph}(\xi)$ and $\hat{\Lambda}^h_{s,sp}(\xi)$ of the approximation on the whole real line. Additionally, there are two eigenvalues located on the spectral gap between the two Fourier symbols, corresponding to the discretization of the homogeneous Dirichlet boundary conditions which, in the DG case, are only weakly enforced. In [29], by means of the multiplier $x \cdot \nabla u$, the stability of the SIPG-n approximations of the Helmholtz equation is proved. This is precisely the multiplier used to prove observability properties for the wave equation in the continuous setting (cf. [44]), but which is not allowed for its conforming finite element approximations due to the fact that the discrete multiplier does not belong to the finite element space. It would be interesting to analyze the discrete observability property by using this multiplier as test function in the DG approximation of the wave equation.
- **DG approximations of the wave equation on nonuniform meshes.** In [54], possible high-frequency pathologies (like the reflection of the discrete characteristic rays before touching the boundary of the space domain) for the FD approximations of the $1-d$ wave equation on smooth nonuniform meshes are identified. In [24], it is proved that, when the grid transformations are strictly concave, the observability from the right endpoint of the space interval holds uniformly as $h$ tends to zero. In [23], the control/observability properties of the mixed finite element approximations of the $1-d$ wave equation on nonuniform meshes are proved to hold uniformly. It would be interesting to use the multiplier and micro-local techniques in [24] and [54] to deal with the propagation properties of the DG approximations of the wave equation. One of the challenges is to analyze the possibility to construct meshes capable to face the pathologies generated by all the critical points of both dispersion relations $\hat{\lambda}^h_{s,ph}(\xi)$ and $\hat{\lambda}^h_{s,sp}(\xi)$.
- **Dispersive properties for the DG approximations of the Schrödinger equation.** In [38], the dispersive properties for the FD schemes of the Schrödinger equation are shown to hold nonuniformly with respect to the mesh size. Also the bi-grid filtering strategy of mesh ratio $1/4$ is proved to be efficient to face these high-frequency pathologies related to the fact that the discrete group

acceleration and group velocity vanish at the wave numbers $\pi/(2h)$ and $\pi/h$, a pathological behavior not happening in the continuous case. However, analyzing these issues for more sophisticated numerical approximations (like the DG ones) is a challenging topic because, in general, the group accelerations of numerical approximations vanish at irrational multiples of $\pi/h$, indicating the necessity of designing and analyzing bi-grid techniques with irrational mesh ratios.

# Appendix A
# Some Technical Proofs

*Proof (of Propositions 4.3 and 4.4 in Chap. 4).* Properties (c1)–(c3) and (d1)–(d3) follow directly from the explicit expressions of the physical and spurious group velocities and elementary calculus tools. Property (c4) follows from the strict positivity of $\hat{\Lambda}^h_{s,sp}(\xi)$ for all $s > 1$ and the equivalent form of $\hat{e}^h_{s,ph}(\xi)$ below:

$$\hat{e}^h_{s,ph}(\xi) = \frac{1}{h^2\hat{\Lambda}^h_{s,sp}(\xi)}\Big\{12(s-1)\Big[(s-1)\Big(3+2\sin^2\Big(\frac{\xi h}{2}\Big)\Big)+2\sin^2\Big(\frac{\xi h}{2}\Big)\Big]$$
$$+\,6(s-\cos(\xi h))\sqrt{\hat{\Delta}^h_s(\xi)}\Big\} > 0.$$

Taking into account the strict positivity of $\hat{\Lambda}^h_{s,sp}(\xi)$ for all $s > 1$ and $\xi \in \Pi^h$, we observe from (4.14) that, for $s \geq 3$, $\hat{e}^h_{s,sp}(\xi)$ is strictly positive, so that (d4) holds for $s \in [3, \infty)$. To conclude (d4), we remark that $\hat{e}^h_{s,sp}(\xi)$ in (4.14) can be written in the following equivalent form:

$$\hat{e}^h_{s,sp}(\xi) = 2s(1-\cos(\xi h)) + \frac{h^2\hat{\Lambda}^h_{s,ph}(\xi)}{2} + \Big(s-\frac{5}{2}\Big)h^2\hat{\Lambda}^h_{s,sp}(\xi).$$

Property (d4) follows by using the fact that $\hat{\Lambda}^h_{s,sp}(\xi) > 0$ and $\hat{\Lambda}^h_{s,ph}(\xi) \geq 0$ for all $\xi \in [0, \pi/h]$ and all $s > 1$ in the above expression of $\hat{e}^h_{s,sp}(\xi)$.

Let us now analyze property (d5). Since $\hat{e}^h_{s,sp}(\xi)$ can be written as

$$\hat{e}^h_{s,sp}(\xi) = \hat{f}^h_{s,sp}(\xi) - (3-s)\sqrt{\hat{\Delta}^h_s(\xi)}, \tag{A.1}$$

with

$$\hat{f}^h_{s,sp}(\xi) := 6(s-\cos(\xi h)) - (3-s)\big(12+2(s-3)(2+\cos(\xi h))\big),$$

A. Marica and E. Zuazua, *Symmetric Discontinuous Galerkin Methods for 1 – D Waves*, SpringerBriefs in Mathematics, DOI 10.1007/978-1-4614-5811-1,

solving $\hat{e}^h_{s,sp}(\xi) = 0$ is equivalent to finding the solutions of

$$\hat{f}^h_{s,sp}(\xi) = (3-s)\sqrt{\hat{\Delta}^h_s(\xi)}. \tag{A.2}$$

Taking into account that, for $s \in (1, 5/2)$, the right-hand side in (A.2) is positive, in order to guarantee the existence of solutions for (A.2), we first have to identify the values of $(\xi, s) \in (0, \pi/h) \times (1, 5/2)$, so that $\hat{f}^h_{s,sp}(\xi) \geq 0$.

We eliminate some values of $s \in (1, 5/2)$ for which there are not solutions of $\hat{f}^h_{s,sp}(\xi) \geq 0$ as a consequence of the monotonicity of $\hat{f}^h_{s,sp}(\xi)$ in $\xi$. First remark that

$$\partial_\xi \hat{f}^h_{s,sp}(\xi) = 2h\sin(\xi h)\big(3-(s-3)^2\big).$$

Depending on the sign of the expression $3-(s-3)^2$, we distinguish two cases:

- For all $s \in (1, 3-\sqrt{3}]$, $\hat{f}^h_{s,sp}(\xi)$ is decreasing, so that

$$\hat{f}^h_{s,sp}(0) = 6(s-1)(s-2) \geq \hat{f}^h_{s,sp}(\xi) \geq \hat{f}^h_{s,sp}(\pi/h) = 2s^2+6s-12$$

  for all $\xi \in (0, \pi/h)$. Then, for all $s \in (1, 3-\sqrt{3}]$, $\hat{e}^h_{s,sp}(\xi)$ in (A.1) is strictly negative, so that we obtain (d5) for $s \in (1, 3-\sqrt{3}] \subset (1, 5/3)$.
- For all $s \in (3-\sqrt{3}, 5/2)$, $\hat{f}^h_{s,sp}(\xi)$ is increasing, so that

$$\hat{f}^h_{s,sp}(0) = 6(s-1)(s-2) \leq \hat{f}^h_{s,sp}(\xi) \leq \hat{f}^h_{s,sp}(\pi/h) = 2s^2+6s-12.$$

  In order to guarantee that $\hat{f}^h_{s,sp}(\xi) \leq 0$, we restrict to $s$ such that $2s^2+6s-12 \leq 0$, i.e., $s \in (3-\sqrt{3}, (\sqrt{33}-3)/2]$. Observe that $(\sqrt{33}-3)/2 < 5/3$, so that so far we can conclude (d5) for $(1, (\sqrt{33}-3)/2] \subset (1, 5/3]$.

  It remains to prove (d5) for $s \in ((\sqrt{33}-3)/2, 5/3]$.

From (A.1), we get that, on the subsets of $\xi \in [0, \pi/h]$ and $s \in ((\sqrt{33}-3)/2, 5/3]$ where $\hat{f}^h_{s,sp}(\xi) \leq 0$, $\hat{e}^h_{s,sp}(\xi)$ is strictly negative. From the above study of the monotonicity of $\hat{f}^h_{s,sp}(\xi)$, we note that $\hat{f}^h_{s,sp}(\xi)$ is nonpositive for all $\xi \in [0, \pi/h]$ when $s \in (1, (\sqrt{33}-3)/2]$ (but this is outside the range of $s \in ((\sqrt{33}-3)/2, 5/3]$ needed in this part of the proof).

For $s \in ((\sqrt{33}-3)/2, 5/3]$, we have that $\hat{f}^h_{s,sp}(0) = 6(s-1)(s-2) < 0$ and $\hat{f}^h_{s,sp}(\pi/h) = 2(s^2+3s-6) > 0$. Due to the strictly increasing character of $\hat{f}^h_{s,sp}(\xi)$, there exists a $\xi^\star \in (0, \pi/h)$ such that $\hat{f}^h_{s,sp}(\xi) \leq 0$ for $\xi \in [0, \xi^\star]$ and $\hat{f}^h_{s,sp}(\xi) > 0$ for $\xi \in (\xi^\star, \pi/h]$. For $\xi \in [0, \xi^\star]$, it is clear from (A.1) that $\hat{e}^h_{s,sp}(\xi)$ is strictly negative.

Let us now focus on the case $\xi \in (\xi^\star, \pi/h]$. Note that $\hat{e}^h_{s,sp}(\xi) < 0$ is equivalent to $\hat{g}^h_{s,sp}(\xi) < 0$, where

$$\hat{g}^h_{s,sp}(\xi) := (\hat{f}^h_{s,sp}(\xi))^2 - (s-3)^2\hat{\Delta}^h_s(\xi).$$

Observe that, for all $s \in ((\sqrt{33}-3)/2, 5/3]$, we have

$$\hat{g}^h_{s,sp}(0) = 36(s-1)^2(2s-5) < 0, \; \hat{g}^h_{s,sp}(\pi/h) = 12(3s-5)(2s^2-3s+3) \leq 0. \tag{A.3}$$

In order to prove (d5) for $s \in ((\sqrt{33}-3)/2, 5/3]$, it is enough to show that $\hat{g}^h_{s,sp}(\xi) < 0$ for all $\xi \in (\xi^\star, \pi/h]$ and $s \in ((\sqrt{33}-3)/2, 5/3]$. More precisely, we will show that $\hat{g}^h_{s,sp}(\xi)$ is strictly increasing on $\xi \in (\xi^\star, \pi/h)$.

It is easy to see that $\partial_\xi \hat{g}^h_{s,sp}(\xi) = 12h\sin(\xi)\hat{f}^h_{s,sp}(\xi)$. The function $\hat{f}^h_{s,sp}(\xi)$ is strictly increasing on $\xi \in [0, \pi/h)$ for all $s \in ((\sqrt{33}-3)/2, 5/3]$. It vanishes at $\xi^* \in (0, \pi/h)$ and $\hat{f}^h_{s,sp}(\xi)$ is negative for $\xi \in [0, \xi^*)$ and positive for $\xi \in (\xi^*, \pi/h]$.

Then $\hat{g}^h_{s,sp}(\xi)$ is strictly increasing for $\xi \in (\xi^*, \pi/h)$ and its maximum value is attained at $\xi = \pi/h$. The proof concludes by observing that

$$\hat{g}^h_{s,sp}(\xi) \leq \hat{g}^h_{s,sp}(\pi/h)$$

and that the right-hand side in the above inequality is nonpositive for all $s \in ((\sqrt{33}-3)/2, 5/3]$ due to (A.3).

In the following, we focus on the property (d6) and on $s \in (5/3, 5/2)$.

By replacing $\cos(\xi h) = X$ in (A.2), we note that $X$ is a solution of the quadratic equation

$$(3-(3-s)^2)X^2 - 2s(2s-3)X + 6s^3 - 22s^2 + 24s - 9 = 0, \tag{A.4}$$

for which the coefficient of the quadratic term $X^2$, $3-(3-s)^2$, is strictly positive. The solutions of (A.4) are explicitly given by

$$X^\pm(s) = \frac{s(2s-3) \pm (s-1)(3-s)\sqrt{6(s-1)}}{3-(3-s)^2}.$$

The values of $s \in (5/3, 5/2)$ for which there exist solutions of (A.2) coincide with those for which at least one of the two numbers $X^\pm(s)$ belongs to $(-1, 1)$. When this happens, the wave number $\xi = \xi_s \in (0, \pi/h)$ in (4.15) can be found as $\xi_s h = \arccos(X^\pm(s))$ and such that $\hat{f}^h_{s,sp}(\xi_s) \geq 0$.

The requirement $-1 < X^\pm(s) < 1$ is equivalent to the following one (in which $\delta(s) := (s-1)(3-s)\sqrt{6(s-1)}$):

$$a(s) := -s^2 - 3s + 6 < \pm\delta(s) < b(s) := -3(s-1)(s-2).$$

There are several possible cases:

- For $s \in S_1 := (2, 5/2)$, both $a(s)$ and $b(s)$ are strictly negative, so that from the two values $\pm\delta(s)$, we choose only the negative one to guarantee $|X^-(s)| < 1$. Both inequalities $-a(s) > \delta(s) > -b(s)$ are satisfied iff $s \in (2, 5/2)$. More precisely, the first one is equivalent to

$$(3s-5)(s^2-6s+6)(-2s^2+3s-3) > 0,$$

which holds for $s \in S_2 := (-\infty, 3-\sqrt{3}) \cup (5/3, 3+\sqrt{3})$, while the second one is equivalent to

$$(2s-5)(s^2-6s+6) > 0,$$

which imposes the restriction $s \in S_3 := (3-\sqrt{3}, 5/2) \cup (3+\sqrt{3}, \infty)$. Finally, we obtain that $|X^-(s)| < 1$ for all $s \in S_1 \cap S_2 \cap S_3 = S_1 = (2, 5/2)$.

- For $s \in S_4 := ((\sqrt{33}-3)/2, 2]$, $a(s) < 0$, while $b(s) \geq 0$. Then, in order to guarantee that one of the inequalities $|X^+(s)| < 1$ or $|X^-(s)| < 1$ holds, we have to determine the additional restrictions on $s$ under which one of the inequalities $-a(s) > \delta(s)$ or $\delta(s) < b(s)$ holds. The first one implies $s \in S_2$ and then $s \in S_2 \cap S_4 = (5/3, 2]$. The second one implies that $s \in S_5 := (-\infty, 3-\sqrt{3}) \cup (5/2, 3+\sqrt{3})$ and then $s \in S_4 \cap S_5 = \emptyset$. Therefore, for all $s \in (5/3, 2]$, we have $|X^-(s)| < 1$.

For all $s \in (5/3, 5/2)$, the unique wave number $\xi_s \in (0, \pi/h)$ defined by (4.15) satisfies the identity (A.2). Moreover, $\hat{f}^h_{s,sp}(\xi_s) = 2(s-1)(3-s)\sqrt{6(s-1)}$ and $\hat{\Delta}^h_s(\xi_s) = 24(s-1)^3$. This concludes the proof of (d6). □

# References

1. Agut C., Diaz J.: Stability analysis of the interior penalty DG method for the wave equation. ESAIM: Math. Model. Numer. Anal. **47**(3), 903–932 (2013)
2. Ainsworth M.: Dispersive and dissipative behaviour of high order DG finite element methods. J. Comput. Phys. **198**(1), 106–130 (2004)
3. Ainsworth M., Monk P., Muniz W.: Dispersive and dissipative properties of DG finite element methods for the second-order wave equation. J. Sci. Comput. **27**(1–3), 5–40 (2006)
4. Antonietti P.F., Buffa A., Perugia I.: DG approximation of the Laplace eigenproblem. Comput. Methods Appl. Mech. Engrg. **195**(25–28), 3483–3505 (2006)
5. Arnold D.N.: An interior penalty finite element method with discontinuous elements. SIAM J. Numer. Anal. **19**, 742–760 (1982)
6. Arnold D.N., Brezzi F., Cockburn B., Marini L.D.: Unified analysis of DG methods for elliptic problems. SIAM J. Numer. Anal. **39**, 1749–1779 (2002)
7. Atkins H.L., Hu F.Q.: Eigensolution analysis of the DG method with nonuniform grids: I. One space dimension. J. Comput. Phys. **182**(2), 516–545 (2002)
8. Bahouri H., Chemin J.Y., Danchin R.: Fourier Analysis and Nonlinear PDEs. Springer, New York (2011)
9. Bardos C., Lebeau G., Rauch J.: Sharp sufficient conditions for the observation, control and stabilization of waves from the boundary. SIAM J. Contr. Optim. **30**, 1024–1065 (1992)
10. Belytschko T., Mullen R.: On dispersive properties of finite element solutions. Modern Problems in Elastic Wave Propagation. Wiley, New York (1978)
11. Brezzi F., Cockburn B., Marini L.D., Süli E.: Stabilization mechanisms in DG finite element methods. J. Comput. Methods Appl. Mech. Engrg. **195**, 3293–3310 (2006)
12. Brillouin L.: Wave Propagation and Group Velocity. Academic Press, New York (1960)
13. Burq N., Gérard P.: Contrôle optimal des equations aux derivées partielles. Cours de l'École Polytechnique (2002)
14. Calo V., Demkowicz L., Gopalakrishnan J., Muga I., Pardo D., Zitelli J.: A class of discontinuous Petrov-Galerkin methods. Part IV: The optimal test norm and time-harmonic wave propagation in $1-d$. J. Comput. Phys. **230**, 2406–2432 (2011)
15. Castro C., Micu S.: Boundary controllability of a linear semi-discrete $1-d$ wave equation derived from a mixed finite element method. Numer. Math. **102**(3), 413–462 (2006)
16. Chironis N.P., Sclater N.: Mechanisms and Mechanical Devices Sourcebook, 3th edn. McGraw-Hill, New York (2001)
17. Cockburn B.: Discontinuous Galerkin methods. Z. Agnew. Math. Mech. **83**(11), 731–754 (2003)

A. Marica and E. Zuazua, *Symmetric Discontinuous Galerkin Methods for 1 – D Waves*, SpringerBriefs in Mathematics, DOI 10.1007/978-1-4614-5811-1, 

18. Cockburn B., Gopalakrishnan J., Lazarov R.: Unified hybridization of DG, mixed and continuous Galerkin methods for second-order elliptic problems. SIAM J. Numer. Anal. **47**(2), 1319–1365 (2009)
19. Cockburn B., Karniadakis G., Shu C.W.: The development of DG methods, in Discontinuous Galerkin methods. Theory, Computation and Applications, Lecture Notes in Computational Science and Engineering, vol. 11, pp. 3–50. Springer, Berlin (2000)
20. Cockburn B., Shu C.W.: The local DG finite element method for convection-diffusion systems. SIAM J. Numer. Anal. **35**, 2440–2463 (1998)
21. Dacorogna B.: Direct Methods in the Calculus of Variations. Springer, Berlin (1989)
22. Demkowicz L., Gopalakrishnan J., Muga I., Zitelli J.: Wavenumber explicit analysis of a DPG method for the multidimensional Helmholtz equation. Comput. Meth. Appl. Mechan. Eng. **213–216**, 126–138 (2012)
23. Ervedoza S.: Observability properties of a semi-discrete $1-d$ wave equation derived from a mixed finite element method on non-uniform meshes. ESAIM: COCV **16**(2), 298–326 (2010)
24. Ervedoza S., Marica A., Zuazua E.: Uniform observability property for discrete waves on strictly concave non-uniform meshes: a multiplier approach, in preparation.
25. Ervedoza S., Zheng C., Zuazua E.: On the observability of time-discrete conservative linear systems. J. Funct. Anal. **254**(12), 3037–3078 (2008)
26. Ervedoza S., Zuazua E.: A systematic method for building smooth controls for smooth data. Discrete Contin. Dyn. Syst. Ser. B **14**(4), 1375–1401 (2010)
27. Ervedoza S., Zuazua E.: The wave equation: control and numerics, in Control of PDEs. Cetraro, Italy 2010, In: Cannarsa P.M and Coron J.M. (eds.), Lecture Notes in Mathematics, vol. 2048, pp. 245–339. Springer, New York (2012)
28. Ervedoza S., Zuazua E.: On the numerical approximation of controls for waves. Springer Briefs in Mathematics. Springer, New York (2013)
29. Feng X., Wu H.: DG methods for the Helmholtz equation with large wave number. SIAM J. Numer. Anal. **47**(4), 2872–2896 (2009)
30. Gérard P., Markowich P.A., Mauser N.J., Poupaud F.: Homogenization limits and Wigner transforms. Comm. Pure Appl. Math. **(50)**(4), 323–379 (1997)
31. Glowinski R.: Ensuring well-posedness by analogy: Stokes problem and boundary control for the wave equation. J. Comput. Phys. **103**(2), 189–221 (1992)
32. Glowinski R., He J., Lions J.L.: Exact and approximate controllability for distributed parameter systems: a numerical approach. Cambridge University Press, Cambridge (2008)
33. Glowinski R., Li C.H., Lions J.L.: A numerical approach to the exact boundary controllability of the wave equation. I. Dirichlet controls: description of the numerical methods. Japan J. Appl. Math. **7**(1), 1–76 (1990)
34. Grenander U., Szegö G.: Toeplitz Forms and Their Applications. Chelsea Publishing, New York (1984)
35. Grote M.J., Schneebeli A., Schötzau D.: DG finite element method for the wave equation. SIAM J. Numer. Anal. **44**(6), 2408–2431 (2006)
36. Hu F.Q., Hussaini M.Y., Rasetarinera P.: An analysis of the DG method for wave propagation problems. J. Comput. Phys. **151**, 921–946 (1999)
37. Ignat L., Zuazua E.: Convergence of a two-grid algorithm for the control of the wave equation. J. Eur. Math. Soc. **11**(2), 351–391 (2009)
38. Ignat L., Zuazua E.: Numerical dispersive schemes for the nonlinear Schrödinger equation. SIAM J. Numer. Anal. **47**(2), 1366–1390 (2009)
39. Infante J.A., Zuazua E.: Boundary observability for the space-discretizations of the one-dimensional wave equation. M2AN, **33**(2), 407–438 (1999)
40. Kalman R.E.: On the general theory of control systems, Proceeding of First International Congress of IFAC, vol. 1, pp. 481–492. Moscow (1960)
41. Lebeau G.:Contrôle de l'équation de Schrödinger. J. Math. Pures Appl. **71**, 267–291 (1992)
42. Lieb E.H., Loss M.: Analysis. Graduate Studies in Mathematics, vol. 14. AMS, Providence, RI (1997)

43. Linares F., Ponce G.: Introduction to Nonlinear Dispersive Equations. Springer, New York (2009)
44. Lions J.L.: Contrôlabilité Exacte, Perturbations et Stabilisation des Systèmes Distribués, vol. 1. Masson, Paris (1988)
45. Loreti P., Mehrenberger M.: An Ingham type proof for a two-grid observability theorem. ESAIM: COCV **14**(3), 604–631 (2008)
46. Macià F.: Propagación y control de vibraciones en medios discretos y continuos. PhD Thesis, Universidad Complutense de Madrid (2002)
47. Macìa F.: Wigner measures in the discrete setting: high frequency analysis of sampling and reconstruction operators. SIAM J. Math. Anal. **36**(2) 347–383 (2004)
48. Markowich P.A., Poupaud F.: The Pseudo-Differential Approach to Finite Difference Revisited, vol. 36, pp. 161–186. Calcolo, Springer, New York (1999)
49. Marica A., Zuazua E.: Localized solutions for the finite difference semi-discretization of the wave equation. C. R. Acad. Sci. Paris Ser. I **348**, 647–652 (2010)
50. Marica A., Zuazua E.: Localized solutions and filtering mechanisms for the DG semi-discretizations of the $1-d$ wave equation. C. R. Acad. Sci. Paris Ser. I **348**, 1087–1092 (2010)
51. Marica A., Zuazua E.: High frequency wave packets for the Schrödinger equation and its numerical approximations. C. R. Acad. Sci. Paris Ser. I **349**, 105–110 (2011)
52. Marica A., Zuazua E.: On the quadratic finite element approximation of one-dimensional waves: propagation, observation, and control. SIAM J. Numer. Anal. **50**(5), 2744–2777 (2012)
53. Marica A., Zuazua E.: On the quadratic finite element approximation of $1-d$ waves: propagation, observation, control and numerical implementation, CFL-80: A Celebration of 80 Years of the Discovery of CFL Condition. Kubrusly C., Moura C.A. (eds.) Springer Proceedings in Mathematics, pp. 75–100 Springer, New York (2012)
54. Marica A., Zuazua E.: Propagation of $1-d$ waves in regular discrete heterogeneous media: a Wigner measure approach, accepted in FoCM
55. Mariegaard J.S.: Numerical approximation of boundary control for the wave equation. Ph.D. Thesis, Department of Mathematics, Technical University of Denmark (2009)
56. Micu S.: Uniform boundary controllability of a semi-discrete $1-d$ wave equation. Numer. Math. **91**(4), 723–768 (2002)
57. Micu S.: Uniform boundary controllability of a semi-discrete $1-d$ wave equation with vanishing viscosity. SIAM J. Control Optim. **47**(6), 2857–2885 (2008)
58. Micu S., Zuazua E.: An Introduction to the controllability of partial differential equations, in Quelques questions de théorie du contrôle. In: Sari T. (ed.) Collection Travaux en Cours, pp. 69–157. Hermann, Paris (2005)
59. Negreanu M., Zuazua E.: Convergence of a multigrid method for the controllability of a $1-d$ wave equation. C. R. Math. Acad. Sci. Paris **338**, 413–418 (2004)
60. Rose H.E.: Linear Algebra. A pure Mathematical Approach. Springer, New York (2002)
61. Sherwin S.:Dispersion analysis of the continuous and DG formulations, in Discontinuous Galerkin methods. Theory, Computation and Applications, Lecture Notes in Computational Science and Engineering, vol. 11, pp. 425–431. Springer, New York (2000)
62. Showalter R.E.: Monotone operators in Banach space and nonlinear partial differential equations. Mathematical Surveys and Monographs, vol. 49. AMS, Providence, RI (1997)
63. Simon J.: Compact sets in the space $L^p(0,T;B)$. Annali di matematica pura ed applicata IV(CXLVI), 65–96 (1987)
64. Sontag E.: Mathematical Control Theory: Deterministic Finite Dimensional Systems. Springer, New York (1998)
65. Stein E.M.: Singular Integrals and Differentiability Properties of Functions. Princeton University Press (1970)
66. Tebou Tcheougoué L.R., Zuazua E.: Uniform boundary stabilization of the finite difference space discretization of the $1-d$ wave equation. Adv. Comput. Math. **26**(1–3), 337–365 (2007)
67. Trefethen L.N.: Group velocity in finite difference schemes. SIAM Rev. **24**(2), 113–136 (1982)

68. Zuazua E.: Exponential decay for the semilinear wave equation with localized damping in unbounded domains. J. Math. Pures et Appl. **70**, 513–529 (1991)
69. Zuazua E.: Boundary observability for the finite difference space semi-discretizations of the $2-d$ wave equation in the square. J. Math. Pures Appl. **78**(5) 523–563 (1999)
70. Zuazua E.: Propagation, observation, control and numerical approximation of waves. SIAM Review **47**(2), 197–243 (2005)
71. Zuazua E.: Controllability and observability of PDEs: Some results and open problems. In: Handbook of Differential Equations: Evolutionary Equations, vol. 3, pp. 527–621. In Dafermos C. M. and Feireisl E. (eds.) Elsevier B. V. North Holland (2007)